Chemical Compound Structures and the Higher Dimension of Molecules:

Emerging Research and Opportunities

Gennadiy Vladimirovich Zhizhin
Russian Academy of Natural Sciences, Russia

A volume in the Advances in
Chemical and Materials Engineering
(ACME) Book Series

IGI Global
DISSEMINATOR OF KNOWLEDGE

Published in the United States of America by
IGI Global
Engineering Science Reference (an imprint of IGI Global)
701 E. Chocolate Avenue
Hershey PA, USA 17033
Tel: 717-533-8845
Fax: 717-533-8661
E-mail: cust@igi-global.com
Web site: http://www.igi-global.com

Library of Congress Cataloging-in-Publication Data

Names: Zhizhin, G. V. (Gennadiæi Vladimirovich), author.
Title: Chemical compound structures and the higher dimension of molecules : emerging research and opportunities / by Gennadiy Vladimirovich Zhizhin.
Description: Hershey PA : Engineering Science Reference, [2018] | Includes bibliographical references.
Identifiers: LCCN 2017028150| ISBN 9781522541080 (hardcover) | ISBN 9781522541097 (ebook)
Subjects: LCSH: Molecular structure. | Chemicals.
Classification: LCC QD461 .Z45 2018 | DDC 541/.22--dc23
LC record available at https://lccn.loc.gov/2017028150

This book is published in the IGI Global book series Advances in Chemical and Materials Engineering (ACME) (ISSN: 2327-5448; eISSN: 2327-5456)

British Cataloguing in Publication Data
A Cataloguing in Publication record for this book is available from the British Library.

All work contributed to this book is new, previously-unpublished material.
The views expressed in this book are those of the authors, but not necessarily of the publisher.

For electronic access to this publication, please contact: eresources@igi-global.com.

Advances in Chemical and Materials Engineering (ACME) Book Series

ISSN:2327-5448
EISSN:2327-5456

Editor-in-Chief: J. Paulo Davim, University of Aveiro, Portugal

MISSION

The cross disciplinary approach of chemical and materials engineering is rapidly growing as it applies to the study of educational, scientific and industrial research activities by solving complex chemical problems using computational techniques and statistical methods.

The **Advances in Chemical and Materials Engineering (ACME) Book Series** provides research on the recent advances throughout computational and statistical methods of analysis and modeling. This series brings together collaboration between chemists, engineers, statisticians, and computer scientists and offers a wealth of knowledge and useful tools to academics, practitioners, and professionals through high quality publications.

COVERAGE

- Metallic Alloys
- Thermo-Chemical Treatments
- Multifuncional and Smart Materials
- Toxicity of Materials
- Fracture Mechanics
- Electrochemical and Corrosion
- Industrial chemistry
- Coatings and surface treatments
- Heat Treatments
- Biomaterials

IGI Global is currently accepting manuscripts for publication within this series. To submit a proposal for a volume in this series, please contact our Acquisition Editors at Acquisitions@igi-global.com or visit: http://www.igi-global.com/publish/.

The Advances in Chemical and Materials Engineering (ACME) Book Series (ISSN 2327-5448) is published by IGI Global, 701 E. Chocolate Avenue, Hershey, PA 17033-1240, USA, www.igi-global.com. This series is composed of titles available for purchase individually; each title is edited to be contextually exclusive from any other title within the series. For pricing and ordering information please visit http://www.igi-global.com/book-series/advances-chemical-materials-engineering/73687. Postmaster: Send all address changes to above address.

Titles in this Series

For a list of additional titles in this series, please visit:
https://www.igi-global.com/book-series/advances-chemical-materials-engineering/73687

3D Printing and Its Impact on the Production of Fully Functional Components Emerging...
Petar Kocovic (Union – Nikola Tesla University, Serbia)
Engineering Science Reference • ©2017 • 115pp • H/C (ISBN: 9781522522898) • US $150.00

Advanced Applications of Supercritical Fluids in Energy Systems
Lin Chen (Tohoku University, Japan & Japan Society for the Promotion of Science (JSPS), Japan) and Yuhiro Iwamoto (Nagoya Institute of Technolog, Japan)
Engineering Science Reference • ©2017 • 682pp • H/C (ISBN: 9781522520474) • US $235.00

Innovative Applications of Mo(W)-Based Catalysts in the Petroleum and Chemical Industry...
Hui Ge (Chinese Academy of Sciences, China) Xingchen Liu (Chinese Academy of Sciences, China) Shanmin Wang (Oak Ridge National Laboratory, USA) Tao Yang (China University of Petroleum, China) and Xiaodong Wen (Synfuels China, China)
Engineering Science Reference • ©2017 • 162pp • H/C (ISBN: 9781522522744) • US $135.00

Sustainable Nanosystems Development, Properties, and Applications
Mihai V. Putz (West University of Timişoara, Romania & Research and Development National Institute for Electrochemistry and Condensed Matter (INCEMC) Timişoara, Romania) and Marius Constantin Mirica (Research and Development National Institute for Electrochemistry and Condensed Matter (INCEMC) Timişoara, Romania)
Engineering Science Reference • ©2017 • 794pp • H/C (ISBN: 9781522504924) • US $245.00

Computational Approaches to Materials Design Theoretical and Practical Aspects
Shubhabrata Datta (Calcutta Institute of Engineering and Management, India) and J. Paulo Davim (University of Aveiro, Portugal)
Engineering Science Reference • ©2016 • 475pp • H/C (ISBN: 9781522502906) • US $215.00

For an enitre list of titles in this series, please visit:
https://www.igi-global.com/book-series/advances-chemical-materials-engineering/73687

IGI Global
DISSEMINATOR OF KNOWLEDGE

701 East Chocolate Avenue, Hershey, PA 17033, USA
Tel: 717-533-8845 x100 • Fax: 717-533-8661
E-Mail: cust@igi-global.com • www.igi-global.com

Table of Contents

Preface

We used to think that the dimension of the molecules is not more than three. However, an analysis of the geometry of molecules using the Euler-Poincaré relation for polytopes of dimensionality n shows that the dimension of many molecules is greater three. Different (often opposite) points of view on space exist long ago. Still Plato, who attached great importance to geometry, suggested that objects of nature, their quality and types of interaction with the environment can be the result of a geometrical structure hidden from us (Yau & Nadis, 2010). A striking example of the influence of hidden geometric structures is string theory, in which, based on the solutions of Einstein's equations, it is postulated that at each point of space a manifold of higher dimension (the Colabi-Yau manifold) is hidden. However, Aristotle considered the space surrounding us to be three-dimensional (Jammer, 1960). With the names of Aristotle and Leibniz, the notion that the real space is inseparably connected with matter is connected. There is no space without matter, just as there is no matter without space: space is the form of the existence of matter (Einstein, 1966). Unfortunately, for many years the opinion, connected with the names of Democritus and Newton, was predominant, that space is the receptacle of all material objects that do not exert any influence on space. This representation has become the basis of Euclidean geometry. According to this representation, the geometric space is continuous, infinite, three-dimensional, homogeneous and isotropic. Opening of Lobachevsky (1945) and Bolyai (1950) non-Euclidean non-contradictory geometry marked the beginning of the rapid development of geometry. Riemann actively continued work on non-Euclidean geometry. In his famous lecture "On the hypotheses underlying the geometry" (Riemann, 1854) he introduces the notion of about n-extended manifold. It is essential that the n-dimensional drawling is determined by Riemann without introducing the infinity of space. Moreover, the infinity of space clearly contradicts Riemann's notion of n-dimensional extension. When it is defined, Riemann initially considers a certain finite domain and, as a sign of the n-dimensionality of the manifold, the position on this manifold is characterized by a change of n one-dimensional extended quantities. Poincare allowed the possibility of the existence of a space with dimension above three. He extended the relations between the numbers of elements of different dimensions in a three-dimensional convex polyhedron into polyhedron of any dimension (polytopes) (Poincare, 1895). Ehrenfest meanwhile, based on the analysis of the stability of the orbits of electrons in an atom, believed that the space was three-dimensional, i.e. there is no atomic structure of matter in a space with a dimension greater three (Ehrenfest, 1920). Later this was confirmed by calculations (Büchel, 1963), as well as by attempts to solve the Schrödinger equation in a space of higher dimension (Gurevich & Mostepanenko, 1971).

In this paper, the molecule is regarded as a convex body. Determining the number of elements of different dimensions entering this body, its dimension is determined by the Euler - Poincaré equation. The existence of a molecule in the form of a convex body of dimension n agrees with Riemann's representations on a finite n-dimensional manifold. In this case, the dimension of the polytope may turn out to be higher than the dimension of the surrounding space, which has the dimensionality 3 under the assumption. Each polytope of dimension n has a boundary complex consisting of elements of smaller dimension. In particular, if the surrounding space has dimension 3, then the boundary elements between the boundary complex and the surrounding space have dimensions 0, 1, 2.

The definition of the dimension of molecules is of great importance in the study of the possibility at their interaction with other molecules. This particular important for living organisms to determine the analysis of disease and treatment, as the geometric condition of complementarity is a key molecule in the interaction. In this study we conducted a systematic analysis of the dimensions of the molecules with the participation of elements of the periodic system (Chapters 1, 2), and we investigate the regularities of filling space with n – dimensional polytopes corresponding to molecules of higher dimension also.

The aim of this study is to prove that virtually very much chemical compounds have the structure of a higher dimension. The results of that research should be published to draw attention to this fact all researches structures of substances. This in turn will give an impulse for the analysis of the interaction of molecules specific substances as objects of higher dimension. This is especially important, for example, in nanomedicine, nanotoxicology and quantum biology.

When constructing a convex polytope corresponding to a molecule, it is first of all necessary to connect the atoms (or functional groups) connected by a covalent bond by the edges. These atoms are identified with the vertices of the polytope. In complex molecules, for example, molecules of supramolecular chemistry or biomolecules, weak (hydrogen) bonds participate in the compounds. Since such bonds can propagate very long distances and are not stable, weak bonds there are not marked by the edges. The edges corresponding to covalent bonds can not form a convex body, as a rule. In order to create a convex polytope it is not necessary to connect vertices by an edge if there is no covalent bond between them. If we connect each vertex from the set of vertices with all the other vertices of this set, then we obtain a polytope, called a simplex. The dimension of the simplex is one less than the number of vertices. However, often, as shown in Chapters 1, 2 of this paper, to form a convex polytope on a set of vertices, it is not necessary to connect each vertex with edges with all other vertices. By observing the homogeneity of the vertices (that is, by an equal number of edges issuing from each vertex), one can reduce the number of edges in comparison with the simplex and obtain a convex polytope of lower dimension on a given set of vertices. Such polytopes can be n-cross - polytope, n - cube, or polytopes without special names.

The fact that these figures are polytopes is proved using the Euler-Poincaré formula, the dimension of the polytope is determined by the same formula. Thus, polytopes of different dimensions can correspond to one the same set of vertices. Physically, the difference in the dimensions of molecules at the same set of vertices (atoms) can be tried to explain, assuming the existence of different energy states of the molecule while maintaining the number of valence bonds of each atom. It should be noted that the space between atoms is not empty. It is characterized by the presence of electrons into it, i.e. of the distribution of electron density (Gillespie & Hargittai, 1991; Sinanoglu, 1965), which is inhomogeneous and has a certain structure. It is obvious that this structure corresponds to the energy state of the molecule and, due to the quantum nature, these states change in a discrete manner. When modeling a molecule by a polytope, the structure of the space inside the molecule can be modeled by the number of edges emanating from each vertex. It is possible to indicate only a finite number of variants of the distribution of edges in a polytope for each given number for condition their homogeneous. It follows from this that the number of possible values of the dimension of a molecule for a given number of vertices is finite and varies discretely.

The structures of biomolecules undoubtedly have the greatest complexity. Conditional images of biomolecules in the form of Fisher and Haworth formulas are distributed, for example, in the form of an "armchair" (Metzler, 1980; Lehninger, 1982). In Chapter 3 new images of molecules of carbohydrates, proteins, nucleic acids in the form of polytopes of higher dimension are presented. This allowed to explain a number of properties of biomolecules, for example, the formation of a chain of carbohydrates in the case of α - *D* glycosidic bond and the absence of a chain of carbohydrates in the case of β - *D* glycosidic bond (Zhizhin, 2016).

A number of serious works on the use of space of higher dimension for the analysis of structure of viruses belong to the Jenner (2006, 2008, 2011, 2016), Indelicato et al. (2012), Twarock, Valiunas, and Zappa (2015).

The form of polytopes corresponding to the structure of molecules of many chemical compounds, as a rule, differs from the form of regular polytopes. Therefore, it is necessary to pay attention to polytopes, in which the conditions for the correctness of the geometric figure are violated. They can have flat faces that are different in form from regular polygons, or the flat faces of a polytope are regular polygons with different number of sides, and so on. Polytopes with such violations of the conditions of correctness can be considered semi-regular if other conditions of correctness are observed. One of the basic conditions for the correctness of a geometric figure is the equality of the vertices of the figure, i.e. compatibility of vertices by movement. In a more general case, we will consider a geometric figure to be semi-regular if its vertices are topologically equivalent, i.e. from each vertex comes an identical number of edges. In Chapter 4, polytopes are studied topologically semi-regular in the above sense.

For the first time, there was announced the existence of convex semi-regular polytopes in four-dimensional space in Gosset's (1900) work. It was a short note, containing only a statement about the existence of three four-dimensional semi-regular finite polytopes with an indication of their composition. Elte (1912) came to the same results. In all the works mentioned, polytopes were considered in which all two-dimensional faces are the same – regular triangles, but the three-dimensional faces in the same polytope can be different. In 1900 and 1910 Stott's work (Stott, 1900, 1910) was published, in which, regardless of Gosset's work, semi-regular polytopes were considered, and the presence of a set of two-dimensional faces in the same polytope of different regular polygons was allowed. It is significant that in all these works no images of some semi-regular polytopes were shown. Images of semi-regular polytopes appeared in print only after 2013 in the works of Zhizhin (2013, 2014, Zhizhin

& Dudea, 2016). Chapter 4 presents the results of these studies, as well as studies related to the expansion of the notion of semi-regular of a polytope in a topological sense. We also consider the hierarchical filling of spaces of higher dimension by regular and semi-regular polytopes. An important example of a semi-regular polytope is the golden hyper-rhombohedron, which, as shown by Zhizhin (2014; see also Zhizhin & Diudea, 2016), is a fundamental region of quasicrystals, i.e., molecules of intermetallics filling the space of the nanoworld. We emphasize that a golden hyper-rhombohedron, having a dimension of 4, fills this space by translation in four directions. The golden hyper-rhombohedron consists from 8 rhombohedrons with dimensional 3 whose faces are rhombus.

As was shown (Zhizhin, 2014), from the lattice of vertices of the golden hyper -rhombohedron in the space 4*D*, one can single out the projections of the golden hyper-rhombohedron from a space of even larger dimension, for example, the space 5*D*. The fundamental domain of this subspace is the product of the gold hyper-rhombohedron 4*D* on the edge, and so on. Pontryagin called structures that are a product of a polyhedron on a one-dimensional segment cylindrical (Pontryagin, 1976). We can say that the product of a polytope on a segment is a prism (Ziegler, 1995) with a base in the form of a polytope. To distinguish it from an ordinary three-dimensional prism, we will call it a polytopic prism. Then the product of one polytope to another polytope is the set of products of one polytope to the set of edges of another polytope.

As noted Ziegler (1995) the products of polytopes are not simplexes, even if the factors are simplexes. In this connection, the developed theory of simplicial polytopes (Pontryagin, 1976; Aleksandrov, 1979) for analysis of the product of polytopes is inapplicable. Therefore, studies of polytope products are of considerable interest. The Chapter 5 and Chapter 7 of the book are devoted to the study of the product of polytopes, which can be called polytopic prismahedrons. It is shown that it is the polytopic prismahedrons that are analogues of the stereohedrons introduced by Delone (Delone, 1961; Delone & Sandakova, 1961), as elements filling without gaps of *n*-dimensional space. It should be noted, however, that principles discrete systems entered by Delone (1937) require consideration. The concept of the discrete points of the system does not require the introduction of a sphere of radius *r* described about anywhere point in the system, within which there are no points (empty balloon introduced into the systems Delone (1937)). The concept of asymptotic decrease in the distance between the points system describes the distribution of points in the diffraction patterns of quasicrystals and corresponds to the scaling process, i.e. continuous change in system scale, open in recent years. Proof of Delone that polytopes in the n - dimensional space, or do not belong to each other, or on the adjacent faces of dimension n - 1 (facets) is not correct, because it uses the concept of three-dimensional space when considering the n-dimensional space. In Chapter 7 there shown that polytopes in the n-dimension space can have common vertices, edges, and common elements of any dimension from 0 to $n - 1$. The Chapter 6 is devoted to study thc polytopcs dual to polytopic prismahedrons. It is shown that the duality to polytopic prismahedrons leads to a new type of polytopes, a poly-incident polytopes that is not previously known. Polytopes in which edges are used simultaneously, incident to different number of elements of greater dimension.

In the Appendix, solutions of the Schrödinger equation in n-dimensional space are investigated. It is shown that the strange statement about the impossibility of the existence of the atomic structure of matter in a space with a dimension above three expressed by Ehrenfest (1920) and confirmed by Büchel (1963) and Gurevich and Mostepanenko (1971) is a consequence of a misunderstanding. Büchel (1963) and Gurevich and Mostepanenko (1971) when analyzing the Schrödinger equation, a potential is introduced in the form of a power function of the radius, and the negative power depends linearly on the dimensionality of the space. Such a representation of the potential function is not supported by any physical considerations, let alone the experimental confirmation of such a function. Such a kind of potential function that leads to the impossibility of quantizing the energy of an atom in a space with a dimension above three. However, if we use an experimentally tested law of the dependence of the potential on the radius (that is, inversely proportional to the radius), then the Schrödinger equation and the higher-dimensional space have a solution with discrete values of quantum numbers.

In the application for the case of the dimensionality of the space equal to 4, an exact solution of the Schrödinger equation is obtained and the quantum numbers corresponding to the space of this dimension are determined.

REFERENCES

Aleksandrov, P. S. (1979). *Selected works. General theory of homology.* Moscow: Science.

Bolyai, Ya. (1950). Appendix. Moscow: State. Ed. Tech.- theor. Lit.

Büchel, W. (1963). Warum hat der Raum drei Dimensionen? *Physikalische Blätter, 1912*(12), 547–549. doi:10.1002/phbl.19630191204

Delone, B. N. (1937). Geometry of positive quadratic forms. *Progresses of Mathematical Sciences, 3*, 16–62.

Delone, B. N. (1961). Proof of the main theorem of the theory of stereohedrons. *Reports. Academy of Sciences of the USSR, 138*(6), 1270–1272.

Delone, B. N., & Sandakova, N. N. (1961). Theory of stereohedrons. *Proceeding Mathematical In. name V.A. Steklov of Academy Sciences of the USSR, 64*, 28–51.

Ehrenfest, P. (1917). Article. *Proceeding Amsterdam Acad., 20*, 200.

Einstein, A. (1966). *The problem of space, field and ether. In Collection of scientific papers* (Vol. 2, pp. 283–286). Moscow: Science.

Elte, E. L. (1912). *The semi-regular polytopes of the hyperspace*. Groningen: University of Groningen.

Euclid, . (2012). *Beginning*. Moscow: URSS.

Gillespie, R. J., & Hargittai, I. (1991). *The VSEPR Model of Molecular Geometry*. London: Allyn & Bacon.

Gosset, T. (1900). On the regular and semi-regular figures in space of *n* dimensions. *Messenger of Mathematics, 29*, 43–48.

Grünboum, B. (1967). *Convex Polytopes*. London: Springer.

Gurevich, L., & Mostepanenko, V. (1971). On the Existence of Atoms in n – Dimensional Space. *Physics Letters. [Part A], 35*(3), 201–202. doi:10.1016/0375-9601(71)90148-4

Indelicato, G., Cermelli, P., Salthouse, D. G., Racca, S., Zanzolto, G., & Twarrock, R. (2012). A crystallographic approach to structural transitions in icosahedral viruses. *Journal of Mathematical Biology, 64*(5), 745–773. doi:10.1007/s00285-011-0425-5 PMID:21611828

Jammer, M. (1960). *Concepts of Space*. New York: Harper and Brothers.

Jenner, A. (2006). Crystallorgaphic structural organization of human rhinovirus serotype 16, 14, 3, 2 and 1A. *Acta Crystallographica. Section A, Foundations of Crystallography, 62*(4), 270–286. doi:10.1107/S010876730601748X PMID:16788267

Jenner, A. (2008). Comparative architecture of octahedral protein cages. II. Interplay between structural elements. *Acta Crystallographica. Section A, Foundations of Crystallography*, *64*(Pt4), 503–512. doi:10.1107/S0108767308012051 PMID:18560167

Jenner, A. (2011). Form, symmetry and packing of biomacromolecules. V. Shell with boundaries at antinodes of resonant vibrations in icosahedral RNA viruses. *Acta Crystallographica. Section A, Foundations of Crystallography*, *67*(Pt6), 521–232. doi:10.1107/S010876731103577X PMID:22011468

Jenner, A. (2016). Symmetry-adapted digital modeling. III. Coarse-grained icosahedral viruses. *Acta Crystallographica. Section A, Foundations of Crystallography*, *72*(Pt3), 324–337. doi:10.1107/S205327331600276X PMID:27126109

Keef, T., & Twarock, R. (2009). Affine extensions of the icosahedral group with applications to the three-dimensional organization of simple viruses. *Journal of Mathematical Biology*, *59*(Pt.3), 287–313. doi:10.1007/s00285-008-0228-5 PMID:18979101

Lehninger, L. L. (1982). Principles of Biochemistry (Vols. 1-3). New York: Worth Publisher, Inc.

Lobachevsky, N. N. (1945). *Geometric research on the theory of parallel lines*. Moscow: The USSR Academy of Sciences Publishing House.

Metzler, D. E. (1980). Biochemistry. The Chemical Reactions of Living Cells (Vols. 1-3). New York: Academic Press.

Poincare A. (1895). Analysis situs. *J. de é Ecole Polytechnique, 1*, 1 – 121.

Pontryagin, L. S. (1976). *Fundamentals of combinatorial topology*. Moscow: Science.

Riemann, B. (1948). *Compositions*. Moscow: OGIZ.

Sinanoglu, O. (Ed.). (1965). Modern Quantum Chemistry (Vols. 1-2). New York: Academic Press.

Stott, A. B. (1900). On certain series of sections of the regular four-dimensional hyper-solids. *Verhandelingen der Koninklijke Akademie van Wetenschappen te Amsterdam*, 3-7.

Stott A. B. (1910). Geometrical deduction of semiregular from regular polytopes and space fillings. *Verhandelingen der Koninklijke Akademie van Wetenschappen te Amsterdam,* 1 – 11.

Twarock, R., Valiunas, M., & Zappa, E. (2015). Orbits of crystallographic embedding of non-crystallographic groups and applications to virology. *Acta Crystallographica. Section A, Foundations of Crystallography*, *71*(Pt6), 569–582. doi:10.1107/S2053273315015326 PMID:26522406

Yau, S.-T., & Nadis, S. (2010). *The Shape of Inner Space. String Theory and Geometry of the Universe`s Hidden Dimensions.* New York: BASIC BOOKS.

Zhizhin, G. V. (2014). *World – 4D.* St. Petersburg: Polytechnic Service.

Zhizhin, G. V. (2016). The structure, topological and functional dimension of biomolecules. *J. Chemoinformatics and Chemical Engineering*, *5*(2), 44–58.

Zhizhin, G. V., & Diudea, M. V. (2016). Space of Nanoworld. In M. V. Putz & M. C. Mirica (Eds.), *Sustainable Nanosystems, Development, Properties, and Applications* (pp. 214–236). New York: IGI Global.

Ziegler, G. M. (1995). *Lectures on Polytopes.* Berlin: Springer Verlag. doi:10.1007/978-1-4613-8431-1

Chapter 1
The Structure and Higher Dimension of Molecules *d*- and *f*-Elements

ABSTRACT

The chapter deals with the chemical compounds formed by the transition elements of the periodic system elements, i.e. d- and f-elements. All these elements are metals and many of them have valuable physical and chemical properties. In the transition elements, the electrons are filled the d- and f-orbital atoms. The filling of the energy levels of the orbitals should occur as the electron energy increases in accordance with the rules of Pauli and Hund. However, many of the transient elements fill electronic orbitals in violation of these rules. This chapter shows that these anomalies can be described by analytic relationships and they lead to an increase in the chemical and physical activity of the elements. It is shown that the molecules of most compounds with the participation of transition elements are of higher dimensionality, which must be taken into account when analyzing their properties.

TRANSITIONAL ELEMENTS IN THE BIOSPHERE

The transitional elements are *d*- and *f*-elements located in the table of Mendeleev between *s* - and *p*-elements (in *s*-elements are completed by electrons *s*-orbitals of the outer level, in *p*-elements are completed *p*-orbitals of the outer level). In the *d*-elements are completed *d*-orbitals of the pre-outer

DOI: 10.4018/978-1-5225-4108-0.ch001

level, in the *f*-elements are completed *f*-orbitals of the pre-outer levels. All these elements form metals. They play a big role in the biosphere - the shell of the Earth, in which there are living organisms (Vernadsky, 2012). They are in the lithosphere, making up the bulk of minerals (Fersman, 1937). In living organisms, due to the biogenic migration of atoms, almost all elements that exist in the crust and water can be detected, including the transitional elements. Transitional elements are centers of biologically active enzymes and hormones, i.e. a small amount in the body, as trace elements, is extremely important for the activity of organisms. However, in the case of excess of the norm (biotic concentration), the transitional elements exhibit toxic properties. The most common transitional element in nature is iron 4.65%, the second element in prevalence is titanium 0, 61%. All the transition elements in order of decreasing their distribution in nature can be represented in the form of a series

Fe (4,65%), Ti (0,62), Mn (0,09%), Zr (0,017%), V (0,015%), Gr (8,3 10^{-3}%), Zn (8 10^{-3}%), Cu (4,7 10^{-3}%), Ce (4,5 10^{-3}%), Co (4 10^{-3}%), Nd (3,7 10^{-3}%), La (1,8 10^{-3}%), Ni (8 10^{-4}%), Th(8 10^{-4}%), Cd (8 10^{-4}%), Sc (6 10^{-4}%), Hf (3,2 10^{-4}%),

U(2,5 10^{-4}%), Ta (2 10^{-4}%), Mo (10^{-4}%), Wo (10^{-4}%), Ag (7 10^{-6}%), Au (5 10^{-6}%),

Hg (5 10^{-6}%), Pt (5 10^{-7}), Y (2,8 10^{-7}%), Rh (10^{-7}%), Re (7 10^{-8}%).

Iron and titanium are constantly in the human body. There are a lot of iron-containing enzymes that catalyze the processes of electron transfer in mitochondria. They are called cytochromes (Metzler, 1980). Cytochromes are ironporphyrins in which all orbitals of the iron ion are occupied by donor atoms of bioligands.

Titanium performs vital functions: catalyzes the synthesis of hemoglobin, increases erythropoiesis and immunogenesis. Titanium compounds accelerate the biosynthesis of amino acids, activate lipoxygenase activity, it increases resistance to various diseases. Titanium compounds are active regulators of free-radical processes and systems for utilization of active forms of oxygen. Other transitional elements also perform important functions in living organisms. For protein, fat and carbohydrate metabolism are necessary Fe, Co, Mn, Zn, Mo, V, B, W. In the synthesis of proteins involved Mg, Mn, Fe,

Co, Cu, Ni, Gr; in hematopoiesis – Co, Ti, Cu, Mn, Ni, Zn; in the breath – Mg, Fe, Cu, Zn, Mn, Co.

All these important functions in living organisms are due to the electronic structure of the outer and pre-outer levels of atoms. This is due to the large number of electrons in the *d*- and *f*-orbitals and, as a consequence, greater possibilities for variations in their number. In addition, there are a large number of free quantum cells on these orbitals that allow the donor-acceptor chemical bond to be realized. All this makes it possible for these atoms to have many valence electrons and to provide a considerable number of coordination bonds in complex compounds.

Energy Levels of Electronic Orbitals of Transitional Elements

It is known that the sequence of filling the electron energy levels and sublevels in many-electron atoms is determined by increasing their energy sequence. The energy of an electron in a many-electron atoms, depending on the principal quantum number *n*, which characterizes the energy level (shell), and orbital quantum number *l*, characterizing the shape of the electron cloud (orbital or sublevel). There are four kinds of forms orbitals *s, p, d, f*. Orbital orientation in space is characterized by the value of the magnetic quantum number m_l. Pointing to the value of the principal quantum number 1, 2, ..., 7, and type of forms of electronic orbitals, it is possible to record the serial number of the experimental increase in energy orbitals located at different energy levels (Gray, 1965)

1s, 2s, 2p, 3s, 3p, 4s, 3d, 4p, 5s, 4d, 5p, 6s, 4f, 5d, 6p, 7s, 5f, 6d, 7p, 6f, 7d. (1)

If to look at the number (1) it can be seen, that the energy of the *d*- orbital of the previous energy level is more energy of the *s* - orbital subsequent energy level $nd > (n+1)s$ (for example, $4d > 5s$). Besides the energy of the *f* - orbitals of the certain energy level is less energy of the *d*- orbital by increasing the number of the energy level on 2 (for example, $4f < 6d$). Consistently filling orbitals at different energy levels, taking into account the capacity of the orbitals (i.e. the (1) presence of each orbital is a certain amount of quantum cells), it is possible in principle to obtain periods, groups and subgroups of the table of chemical elements of Mendeleev (1934). When filling out the orbital electrons it is recommended to use a number (1), as well as the Pauli's principle (Pauli, 1925) and Hund's rule (Hund, 1927).

According to the Pauli's principle in quantum cell cannot be more than two electrons with opposite spins (spin quantum number m_s equal to ± 1/2 respectively). The atom cannot has two electrons with the same values of the four quantum numbers. The number of possible quantum cells at this sublayer (orbital) is $m_l = 2l + 1$. The maximum number of electrons in that sublevel equal to $2m_l$. The number of possible quantum cells at any level is equal to n^2, and the maximum number of electrons at this level is $2n^2$.

According with Hund's rule the filling of the orbitals it is at first by one electron in each quantum cell with the same orientation of the spins, and only after this the quantum cells it are filled with second electrons with opposite spins. Thus, the total spin orbitals must be maximized.

Filling the orbitals on the Pauli principle and Hund's rule corresponds to the ground state of the atom with the lowest energy. When atom reporting additional energy than one or more electrons in the atom move to a higher energy level. In this case, the atom to become excited.

However, for increasing the number of electrons in the atom among of the chemical elements appearing elements in order to fill electron orbitals are the deviations at a number (1.1) and Hund's rule. It is assumed that these anomalous cases are not significant as a whole for the entire table of chemical elements (Arkel, 1931; Karapet`yants & Drakin, 1994). However, a detailed analysis of the electronic formulas of chemical elements table of Mendeleev shows that these anomalous elements are many and it is important that in their number reach elements with exceptional properties such as chromium, platinum, gold, silver, uranium, and others. In the light of the opening of all the new elements with many electrons is of interest to 1) organize the anomalies in the filling of the atomic electron orbitals, 2) try to find patterns in these anomaly filled electron orbitals, and 3) determine the characteristics of the compounds of anomalous elements.

Anomalies in Filling the Electronic Orbitals and Their Analysis

For clarity, the series (1) can represented graphically depending on the principal quantum number *n* (Figure 1).

However, taking into account that the electron energy *E* depends on two quantum numbers it is useful to the further analyze to present this dependence on two coordinates:the principal quantum number *n* and orbital quantum number *l*. So the orbitals *s, p, d, f* correspond to the values of the orbital

Figure 1. Distribution of energy E orbitals s, p, d, f of the atoms, depending on the principal quantum number n

quantum number l = 0, 1, 2, 3, reconstruct Figure 1 as a function *E (n, l)* (Figure 2).

Figure 2 shows that the energy of the electron *E* increases with increasing *n* as well as *l*, but do the nature of these relationships is different from linear. Comparing the current data on the electronic formulas of chemical elements (Gray, 1965) with a number (1), you can make sure that all cases elements with anomalous filling of the atomic electron orbitals are *d*- and *f*-elements. Moreover, from the total number of elements (68) near their third (21) are anomalous elements. In Table 1 and Table 2 shows the structure of the outer shell, built on a number of energy (1), and the actual structure of the outer electron shells, respectively, *d*- and *f*-elements that have anomalies in the order of filling of electron orbitals. We consider first an anomalous element in the table 1, i.e. chromium element. Situated in the table of the elements after vanadium, having in the outer shell electrons $3d^3 4s^2$, chrome atom differs from vanadium atom the one additional electron. This electron, according to

Figure 2. Distribution of energy E orbitals s, p, d, f of the atoms, depending on the principal quantum number n and orbital quantum number l

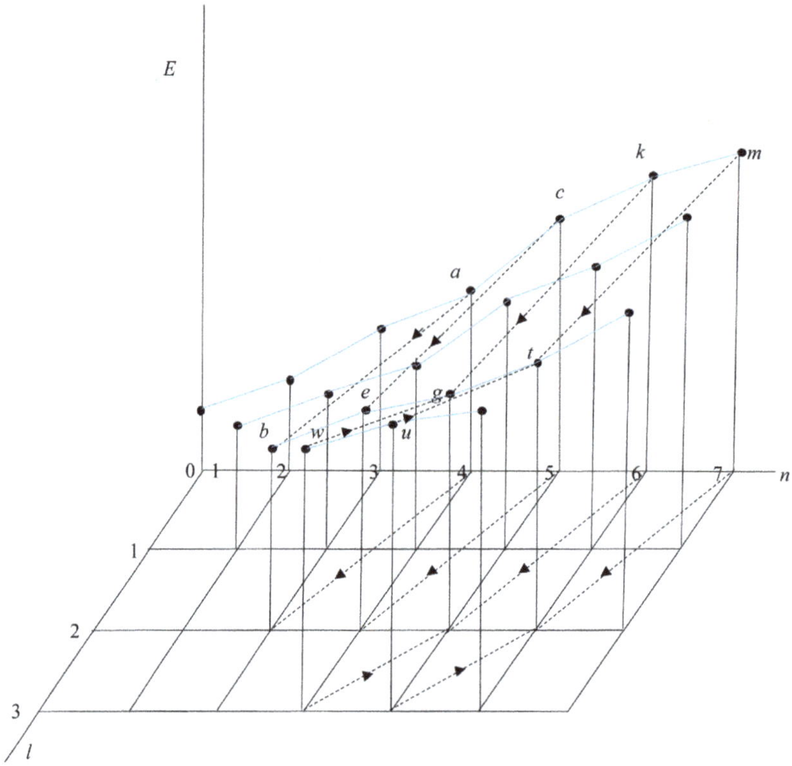

the number of the energy (1) and Hund's rule must be to act on *d*- orbital of the third energy level. He there and goes, but one electron with *s*-orbital on the fourth energy level goes on *d*- orbital of the third energy level despite the Hund's rule governing the sequential filling of quantum cells at this energy level. While at the same time in line with other approved of rules Hund atom acquires the maximum value of the total spin of the shell. This is points to the inconsistency Hund's rule. As a result, the electronic formula of chromium atom is $3d^5 4s^1$ instead $3d^4 4s^2$, calculated on a number (1). Due to the transition of an electron from 4*s*-orbital in the 3*d*-orbital energy of the atom to become higher than the lowest possible value of the energy. We can assume that the chromium atom in its initial position without outside influence is already excited. Figure 2 shows the transition of an electron in the atom of chromium from the 4*s*-orbital on 3*d*-orbitals can be represented by a dotted line with an arrow pointing from point *a* to point *b*.

On the flat (n, l) projection of this line is straight connection point (n = 4, l = 0) with point (n = 3, l = 2).

The same line (a, b) and its projection meets the chemical element Cu. Here also contrary to a number of energy (1) and Hund's rule a one electron goes from the 4s - orbital in the 3d- orbital filling it completely. In so doing the value of the total spin in comparison with the formula for a number of energy $3d^9 4s^2$ is not changed. Diu to of transition electron with 4s-orbital on 3d-orbital having more energy copper atom in its initial state is already excited.

In the atoms of elements Nb, Mo, Ru, Rh, Pd, Ag contrary to a number of energy (1) and Hund's rule electrons go with 5s-orbitals on 4d-orbital with higher energy. In Figure 2 this chemical elements corresponds to the dotted line with an arrow from point c to point e. Its projection on the plane (n, l) of the connecting points (n = 5, l = 0) and (n = 4, l = 2). These atoms as compared with a number of energy (1) are an excited, in the atom of palladium from the 5s - orbital going to 4d – orbital all at once two electrons. Platinum and gold in Figure 2 corresponds to the dotted line with an arrow connecting points (n = 6, l = 0) and (n = 5, l = 2).

The latter anomalous s - element in Table 1, i.e., element Rg corresponds to Figure 2 the dotted line with an arrow connecting points m and t, and its projection onto the plane (n, l) joining the points (n = 7, l = 0), (n = 6, l = 2). They describe the transition of an electron from 7s-orbital to 6d- orbital.

From Figure 2, it follows that the electron transitions in anomalous d-elements describes the straight line in the plane (n, l) through two points (n_0, l_0) and $(n_0 - 1, l_0 + 2)$, where n_0, l_0 are coordinates at the beginning of the transition point.

Considering that $l_0 = 0$, the equation of this line is given by $l = 2(n_0 - n)$, $n_0 = 4 \div 7$. In the final state $n = n_0 - 1$, $l = 2$. Therefore, the sum of the principal quantum number and the orbital quantum number at the end of the transition is equal to $n_0 + l_0 + 1$, i.e. on one greater than the sum of those numbers at the beginning of the transition.

Thus, it is possible to formulate a general rule for the electron transitions in anomalous d-elements: in anomalous d-elements is the transition of one or two electrons from the ns-orbital on the $(n - 2)$ d-orbital, ($n = 4 \div 7$) with higher energy. The sum of the principal quantum number and the orbital quantum number after the transition is increased by 1.

Let us now consider anomalous f-elements. The first anomalous f-element in Table 2 is lanthanum. He is in the table of the chemical elements is preceded

Table 1. Anomalous d- elements

Number of Element	Symbol of Element	The Structure of the Outer Shell on a Number of Energy	The Actual Structure of the Outer Shells
24	Cr	$3d^4 4s^2$	$3d^5 4s^1$
29	Cu	$3d^9 4s^2$	$3d^{10} 4s^1$
41	Nb	$4d^3 5s^2$	$4d^4 5s^1$
42	Mo	$4d^4 5s^2$	$4d^5 5s^1$
44	Ru	$4d^6 5s^2$	$4d^7 5s^1$
45	Rh	$4d^7 5s^2$	$4d^8 5s^1$
46	Pd	$4d^8 5s^2$	$4d^{10} 5s^0$
47	Ag	$4d^9 5s^2$	$4d^{10} 5s^1$
78	Pt	$5d^8 6s^2$	$5d^9 6s^1$
79	Au	$5d^9 6s^2$	$5d^{10} 6s^1$
111	Rg	$6d^9 7s^2$	$6d^{10} 7s^1$

Table 2. Anomalous f-elements

Number of Element	Symbol of Element	The Structure of the Outer Shell on a Number of Energy	The Actual Structure of the Outer Shells
57	La	$4f^1 6s^2$	$5d^1 6s^2$
58	Ce	$4f^2 6s^2$	$4f^1 5d^1 6s^2$
64	Gd	$4f^8 6s^2$	$4f^7 5d^1 6s^2$
89	Ac	$5f^1 7s^2$	$6d^1 7s^2$
90	Th	$5f^2 7s^2$	$6d^2 7s^2$
91	Pa	$5f^3 7s^2$	$5f^2 6d^1 7s^2$
92	U	$5f^4 7s^2$	$5f^3 6d^1 7s^2$
93	Np	$5f^5 7s^2$	$5f^4 6d^1 7s^2$
96	Cm	$5f^8 7s^2$	$5f^7 6d^1 7s^2$
97	Bk	$5f^9 7s^2$	$5f^8 6d^1 7s^2$

by barium, which has on the outer shell electrons $6s^2$. Lanthanum atom differs from barium atoms with one extra electron. In accordance with a number of energy (1) the electron must go on 4*f*-orbital, but it actually enters the free orbital to a higher energy $5d^1$. Figure 2 it is represented by the dashed line with an arrow connecting the points *w* and *g*. There is the total spin is not changed and remains minimal. The projection of this line onto a plane (*n, l*) connecting a point ($n = 4$, $l = 3$), ($n = 5$, $l = 2$). This same line and its projection corresponds transition of an electrons in the elements cerium and gadolinium. In elements Ac, Th, Pa, U, Np, Cm, Bk electrons proceeds with 5*f*-orbitals to 6*d* -orbital at a higher energy. These transitions in Figure 2 corresponds to the dotted line with an arrow connecting a point *u* and point *t*. The projection of this line onto a plane (*n, l*) unchanged. Thus, it is possible to formulate a general rule of transitions electrons in anomalous *f*-elements. In the anomalous *f*-elements is the transition of one electron with *n f*-orbital to the $(n + 1)$ *d*-orbital ($n = 4$; 5) with higher energy. The sum of the principal quantum number and the orbital quantum number remains in the transition unchanged.

A common characteristic of anomalous elements is that in its electrons with *s*- and *f*-orbitals over-tighten on the *d*-orbitals with increasing energy of the atoms (i.e., we can assume that they are in an excited state). In number of work trying to explain the existence of anomalous elements (with reference to the Hund's rule) resistance *d*-orbitals in half or completely filled with electrons (Arkel, 1931; Karapet`yants & Drakin, 1994). However, there are the many of anomalous elements in which the *d*-orbitals of the outer shells not filled halfway or completely by electrons. Such anomalous elements are Nb, Ru, Rh, Pt, La, Ce, Gd, Ac, Th, Pa, U, Np, Cm, Bk. In addition, a statement about the stability of half or fully filled orbitals is not confirmed. To all appearance there are other principles expressed in this paper of empirical rules of thumb transition of electrons in anomalous elements.

If the periods of the table of chemical elements arranged in a line, as it was at Mendeleev (1934), we find an interesting pattern (Zhizhin, 1998) - anomalous elements form three groups of elements sharing the rest of the set of elements into four parts. The first group comprises 10 elements Cr, Nb, Mo, La, Ce, Ac, Th, Pa, U, Np. This group can conditionally called a chromium group. Almost all the elements of this group are durable metals. They actively manifest themselves as catalysts. Moreover, *d*- the elements of the group (Cr, Nb, Mo) are characterized by a half or nearly half-filled *d*-orbitals on atoms subshell and *f*-elements from the group (La, Ce, Ac, Th, Pa, U, Np) characterized by the presence of one electron (maximum of two) to *d*-orbital subshell. The second group includes eight elements Cu, Ru, Pd, Ag, Rh, Gd, Cm, Bk. It also metals, but they are mostly mild. You can call them a group of copper. They are also active as catalysts. Moreover, *d*-elements of the group (Cu, Ru, Pd, Ag, Rh) are characterized by full (or nearly full) filling *d*- orbital subshell, and *f*-elements from this group (Gd, Cm, Bk) are characterized by the presence of a single electron in the *d*-orbital subshell.

The third group includes 3 *d*-element Pt, Au, Rg. Mention may be made of platinum group or a group of noble metals.

The elements characterized by full (or nearly full) filling *d*-orbitals of the subshell.

In connection with the discovery of new elements and the completion of the seventh periodic table of Mendeleev and the possible opening in the future of new elements it is possible to predict the existence of new anomalous elements of the eighth period with the new unusual properties.

CHEMICAL COMPOUNDS OF ANOMALOUS ELEMENTS

The desire of valence electrons to increase the number of *d*-electrons should lead to some features of the compounds of anomalous elements. We consider which the features. We choose to consider the one *s*-element of the three mentioned groups of anomalous elements. The first anomalous element in the chromium group is Cr. It in consequence of the anomalies have one valence electron on the 4 *s*-orbital and five of the electrons on 3*d*-orbital. This allows have of chromium a valence equal 6 in many compounds. Since crystalline chromium oxide CrO_3 consists of chains of tetrahedrons CrO_4, united in two vertices. Each tetrahedron has located in the center of an atom of chromium associated by double bond with each of the four oxygen atoms at the vertices of a tetrahedron. All molecules have the form of a tetrahedron with the center, as shown in (Zhizhin & Diudea, 2016; Zhizhin, Khalaj & Diudea, 2016), have a dimension of 4, i.e., crystalline chromium oxide is a chain of polytopes of dimension 4, united in two vertices (Figure 3). If instead of double bonds are one-time connection with the chromium atom monovalent groups (such as hydroxyl groups), the molecule will be the center of the octahedron, which also would have dimension 4.

Figure 3. The chain of molecules CrO_3

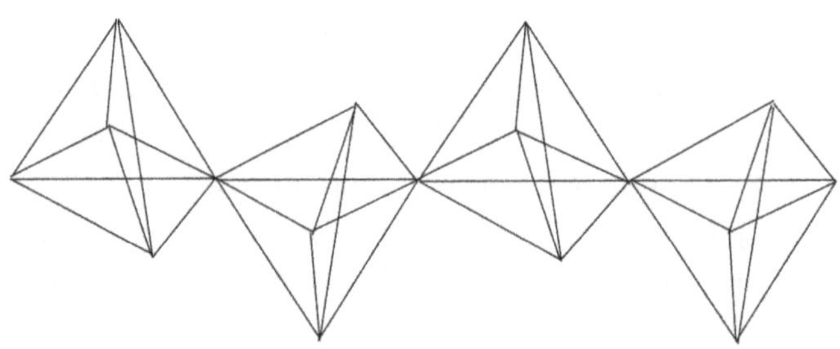

There are even more complex chromium compounds (see Gillespie, 1972; Gillespie & Hargittai, 1991), the dimension of molecules, which is more than four. It is clear that all the other anomalous elements with half (or nearly half) filled *d*-orbitals of the subshell will have similar compounds having a molecular of higher dimensions.

If the anomalous elements have one electron in the outer orbital *s* and subshell *d* completely (or almost completely) filled, then the element at the expense of *s* electron forms a linear molecule, such as a linear molecule oxide X_2O, where X is the anomalous element (Cu, Pd, Ag, Pt, Au, Rg). However, due to the donor-acceptor bond linear molecule can form complex structures in the space. We choose element Cu from second group of anomalous elements. Figure 4 shows an exemplary structure formed by linear molecules Cu_2O

Each oxygen atom in the structure of Figure 4 bonded to four metal atoms (Cu). Two covalent bonds due to the formation of electron pairs divided: one *s* - electron metal atom and a *p* - electron atom of oxygen. In addition, there are two more donor-acceptor chemical bond due to the transfer of two electrons from the *s* - orbital and two electrons from the *p* - orbitals of the

Figure 4. The structure of the compound Cu_2O
A black small circle is oxygen atom. A brown circle is copper atom.

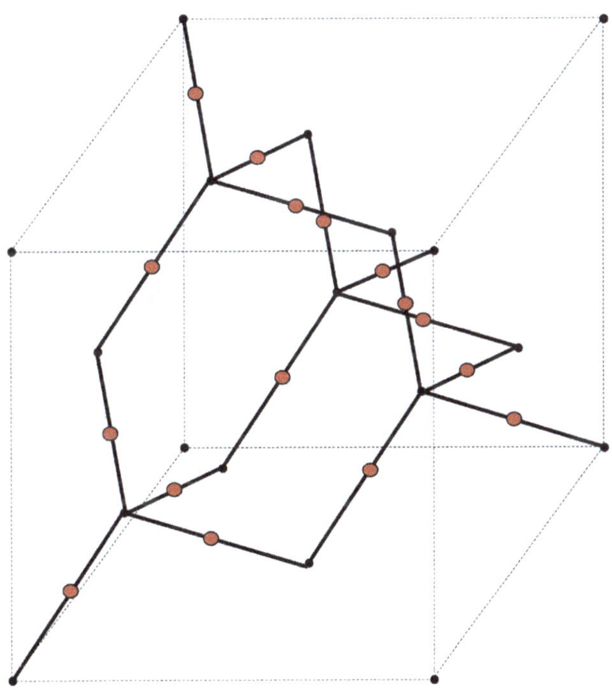

oxygen atom to vacant quantum cell of orbital metal. Thus, oxygen atom acquires valence equal four. In addition, each metal atom is linearly between two oxygen atoms.

In this structure, oxygen atoms (except oxygen atoms located at the vertices of a cube) form structure is topologically equivalent to the structure of carbon atoms in the molecule of adamantane. As shown in the article of Zhizhin (2014a) on the basis the monograph Zhizhin (2014b), the dimension of this molecule is 4. However, the two molecules comprising 10 oxygen atoms have free unallocated space. Therefore, if we set the task of finding the unit cell structure of copper oxide without filling cracks and gaps to help translation the entire space, we need to build politopic prismahedron Zhizhin (2015), with bases in the form polytopes corresponding to these molecules. Taking a line segment equal to the length of the edge of the cube, in which is inscribed the structure including 10 oxygen atoms, multiply the polytope corresponding to this structure for this segment. We obtain politopic prismahedron of dimension 5. With this politopic prismaehdron can fill space without gaps and clearances.

From the third group of anomalous elements we choose gold (Au). The outer shell of gold atom has one 6*s* - electron and a completely filled 5*d*-orbital. In the compound chlorine triphenylphosphine of gold $(Ph_3P)AuCl$ gold atom, giving one *s* - electron to chlorine atom, forms ion $Au(Ph_3P)_3^+$ with trigonal coordination (Gillespie, 1972; Perrin, Armarego & Perrin, 1980). Following Zhizhin (2016) we denoted phosphine molecule as a functional group of the compound.

Then ion $Au(Ph_3P)_3^+$ represented in the form of three tetrahedrons with the center, having one common vertex - a gold atom. In the center of each tetrahedron is located phosphorus atom and the remaining vertices of the tetrahedrons are occupied introduced functional groups Ph_3 (Figure 5).

The functional dimension of each tetrahedron with center is still equal to 4. Thus, the ion $Au(Ph_3P)_3^+$ is a collection of three polytopes of dimension 4, having a common vertex. The assertion is proven

Theorem 1

The ion Au $(Ph_3P)_3^+$ has dimension 5.

Figure 5. Ion $Au(Ph_3P)_3^+$
A white circle is gold atom. A black small circle is phosphorus atom. A black big circle is functional groups Ph_3

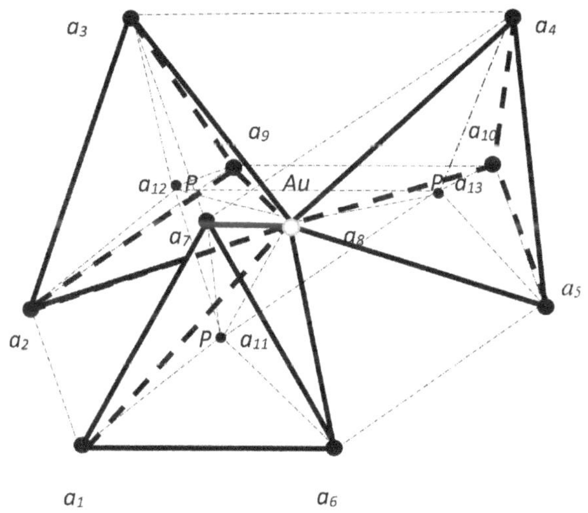

Proof

To prove the necessity of the three tetrahedrons with a common vertex to form a convex shape. Connect the vertices of a_3, a_4, a_7 by line segments, forming a triangle $a_3\ a_4\ a_7$. Connect also the three centers of the tetrahedrons with each other, forming a triangle $a_7 a_{11} a_{13}$. Connect the center of the tetrahedrons with vertices corresponding of the tetrahedrons and vertices in the grounds of the tetrahedrons, forming a hexagon $a_1\ a_2\ a_9\ a_{10}\ a_5\ a_6$ (thin lines on Figure 5). Define dimension polytope in Figure 5 on the Euler- Poincare equation (Poincare, 1895)

$$\sum_{i=0}^{n-1}(-1)^i f_i(P) = 1-(-1)^i, \quad (2)$$

f_i is the number of the elements with the dimension i at polytope P; n is dimension of the polytope P.

To calculate the number of elements of large dimensions we turn first to a simple polytope, a part of a polytope in Figure 5. Temporarily excluded from Figure 5 the centers of the tetrahedrons – a_{11}, a_{12}, a_{13}, and all edges emanating from these vertices. Then, the polytope has 13 vertices, i. e. $f_0 = 10$.

The number of edges of the polytope is sum the number edges of three tetrahedrons ($6{\bullet}3 = 18$), the number of edges connecting tetrahedrons at the base figure (3), the number of the edges connecting vertices of the tetrahedrons at top of Figures 5. Thus, the number of edges polytope without centers of the tetrahedrons equal 24, i.e. $f_1 = 24$ . . The number of flat elements is sum of the flat faces of tetrahedrons ($4{\bullet}3 = 12$), 1 hexagon, 3 triangles between tetrahedrons at base of figure, 3 lateral tetragons, 4 triangles of tetrahedron at top of figure. Thus, the number of flat elements is 23, i.e., $f_2 = 23$ . . The number of three-dimension elements is sum of 4 tetrahedrons, 3 figure between tetrahedrons, 1 hexagon at base and figure composed from boundary flat faces. Thus, the number of three-dimension elements is 9, i.e. $f_3 = 9$. Substituting the values f_i in equation (2), we see that it holds for $n = 4$.

$$10 - 24 + 23 - 9 = 0.$$

Therefore, three tetrahedrons with common vertices is polytope with dimension 4.

For add centers in tetrahedrons the number of the vertices becomes equal 13, i.e. $(f_0)_c = 13$. For this there add the number of the edges: $4{\bullet}3 = 12$ edges in tetrahedrons with centers, and 3 edges connecting centers. Thus, common number of edges on Figure 5 equal 39, i.e. $(f_1)_c = 39$. The number flat faces there increases on 18 triangles in the tetrahedrons, 4 triangles in tetrahedron $a_{11}a_{12}a_{13}a_8$, 6 tetragons with vertices part which are centers of the tetrahedrons. Thus, common number of flat faces on Figure 5 equal 51, i.e. $(f_2)_c = 51$. For add centers the number of three-dimensions faces increases on $4{\bullet}3 = 12$ tetrahedrons into tetrahedrons with centers, tetrahedron $a_{11}a_{12}a_{13}a_8$, figure a_1 a_2 a_9 a_{10} a_5 a_6 $a_{11}a_{12}a_{13}a_8$, prism $a_{11}a_{12}a_{13}a_3a_4a_7$, 3 pyramids with vertex a_8 ($a_8a_1a_2a_{12}a_{11}$, $a_8a_5a_6a_{11}a_{13}$, $a_8a_9a_{10}a_{12}a_{13}$), 3 prism ($a_1a_2a_3a_7a_{11}a_{12}$, $a_5a_6a_4a_7a_{11}a_{13}$, $a_9a_{10}a_3a_4a_{12}a_{13}$). Thus, common number of three-dimension faces on Figure 5 equal 30, i.e. $(f_3)_c = 30$.

It is known from the preceding that the Figure 5 has polytopes of dimension 4. Each tetrahedron with center there is polytope of dimension 4 and 3 tetrahedrons without center, but with a common vertex, there is a polytope of dimension 4. In addition, in Figure 5 between any two tetrahedrons with the center is polytope dimension 4. Obviously, such polytopes are 3. That to proof this statement we consider any polytope from them. For example, the polytope $a_4a_5a_6a_7a_{11}a_{13}$ (Figure 6).

Figure 6. The polytope $a_4a_5a_6a_7a_{11}a_{13}$

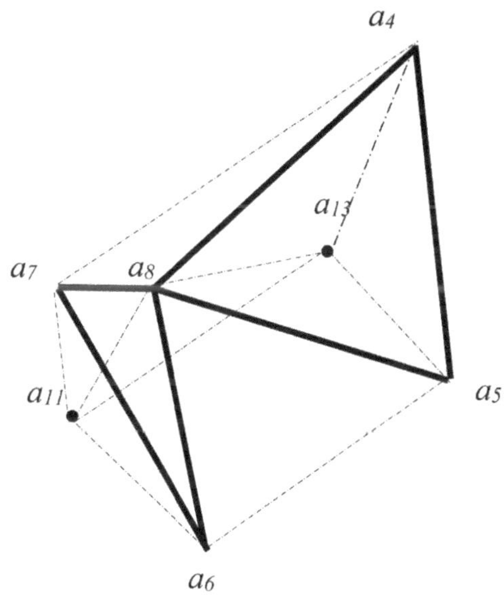

Figure 6 has 7 vertices $(f_0 = 7)$; 15 edges $a_{11}a_7$, $a_{11}a_8$, $a_{11}a_6$, $a_{11}a_{13}$, a_7a_6, a_7a_8, a_7a_4, a_8a_4, a_8a_{13}, a_8a_5, a_6a_5, $a_{13}a_5$, a_4a_{13}, a_4a_5, a_8a_6; 14 flat faces $a_{11}a_8a_6$, $a_{11}a_7a_8$, $a_6a_7a_8$, $a_{11}a_7a_6$, $a_8a_5a_{13}$, $a_8a_{13}a_4$, $a_5a_{13}a_4$, $a_8a_5a_4$, $a_6a_8a_5$, $a_{11}a_8a_{13}$, $a_7a_8a_4$, $a_6a_{11}a_{13}a_5$, $a_6a_5a_7a_4$, $a_{11}a_7a_{13}a_4$; 6 three-dimension faces $a_6a_{11}a_8a_7$, $a_8a_5a_{13}a_4$, $a_7a_8a_{11}a_{13}a_4$, $a_6a_{11}a_8a_{13}a_5$, $a_{11}a_6a_7a_{13}a_5a_4$, $a_6a_7a_8a_4a_5$. Therefore, for Figure 6 are $f_0 = 7, f_1 = 15, f_2 = 14, f_3 = 6$. Substituting the values f_i in equation (2), we see that it holds for $n = 4$

$$7 - 15 + 14 - 6 = 0.$$

This proof that Figure 6 has dimension 4.

As each figures in polytope on Figure 5 is 3, so common number polytopes with dimension 4 in Figure 5 equal 7, i.e. $(f_4)_c = 7$. Substituting the values $(f_i)_c$ in equation (2), we see that it holds for $n = 5$

$$13 - 39 + 51 - 30 + 7 = 2.$$

Theorem is proved.

A BINARY CHEMICAL COMPOUNDS WITH TRANSITION ELEMENTS AND THEIR STRUCTURE

The various binary chemical compounds have a limited number of typical structures (Fersman, 1937). In this section, the structures of binary compounds involving transition elements is considered. The simplest compound of transition elements with three-dimensional structure has a cubic unit cell, for example, oxides of transition elements. Since such structures form chlorides, bromides and iodides of alkali metals, we will refer to these structures rock salt structures (Table 3). Many binary chemical compounds have the structure as adamantane molecule. In work of Zhizhin (2014a) it was proved that the adamantane molecule consisting of 10 carbon atoms that make up the bulk of the unit cell of the diamond has of dimension 4. In compounds Ag_2O, Cu_2O at locations 10 of the carbon atoms are oxygen atoms, and atoms of copper and silver are arranged linearly between oxygen atoms. Many drugs are also a group of 10 carbon atoms as in the adamantane molecule. Among the inorganic and organometallic compounds have a number of structural analogues of adamantine (Table 3). All of these compounds have dimension 4 or even higher.

A series of binary compounds have a structure in the form of cube with centrum as in titanium chloride at which titanium ions are arranged in the centrum of the cube bat chlorine ions are arranged in vertices of the cube. We will call these structures titanium chloride structure. Haw it is shown in work of Zhizhin and Diudea (2016) this structure has dimension 4.

A series of binary compound have a structure of the mineral rutile TiO_2. In this compound each titanium atom is surrounded by six the oxygen atoms in the octahedral coordination. To compounds with such structure to concern for example fluorides of copper, zinc, magnesium, manganese, cobalt, nickel. We will call these structures rutile structure. In work of Zhizhin and Diudea (2016) it is shown that octahedron with centrum have dimension 4. Therefore, all these structures have dimension 4. A series of binary compounds have structure of wurtzite - mineral ZnS, in which from compound ZnS with the structure of the adamantane zinc atom and sulfur atom have the tetrahedron coordination. The centrum each tetrahedron is vertex of another tetrahedron. The wurtzite structure have compounds ZnO, CdS, ZnS. The dimension of this structure remains unknown. It will be defined in the next section. A series of binary compounds have a fluorite structure – mineral CaF_2 (fluorspar). Each calcium ion in this structure is in cube surrounded by fluorine ions, and each fluorine ion is in tetrahedron surrounded by calcium ions. The fluorite structure have for example chlorides of transition metals (Table 3). The dimension of this compounds it will be defined also in the one next sections.

The Dimension of the Wurtzite

In the structure of wurtzite every atom of one component has a tetrahedral environment of the atoms of the other component. This results to arrangement of tetrahedrons with center so that vertex one tetrahedron is the center of another tetrahedron (Figure 7).

Table 3. Binary compounds of the transition elements

N	Type of the Structure	The Compounds Transition Elements With This Type of the Structure
1	rock salt	MnO, FeO, CoO, NiO, CdO
2	Adamantane	Ag_2O, Cu_2O, ZnS, CuCl
3	titanium chloride	TiCl
4	Rutile	MnF_2, CoF_2, NiF_2, CuF_2, ZnF_2, MnO_2, MoO_2
5	Wurtzite	ZnO, CdO, ZnS
6	Fluorite	CdF_2, $MnCl_2$, $FeCl_2$, $CoCl_2$, $NiCl_2$, $ZnCl_2$, $CdCl_2$, Cu_2S

If to carry construction of atoms in Figure 7 on this principle, we obtain a spatial lattice, the unit cell the lattice is a convex shape it is shown in Figure 8.

This figure is the unit cell structure of the wurtzite. In Figure 8, solid lines represent chemical bonds of the atoms, and the dotted lines are only geometric sense outlining contours of the figure. We define the dimension of this figure by the Euler-Poincare equation (2). The number of vertices of this figure is equal to 14, i.e., $f_0 = 14$. The number of edges is equal to 29, i.e., $f_1 = 29$. The number of two-dimensional faces is the sum of the number of triangles (8) and number of quadrangles (13), i.e., $f_2 = 21$. The number of three-dimensional faces is equal to 6. This figures are *abcghkon, gceruo, cefump, cdfulm, bcdklu,* and all shape on Figure 8 without inner partitions, i.e., $f_3 = 6$. Substituting these values $f_i, (\mathrm{i} = 0,1,2,3)$ in the Euler-Poincare equation (2), we find that it is satisfied for $n = 4$.

Figure 7. The tetrahedral coordination atoms in wurtzite
A white circle is atom of one component. A black circle is atom of other component

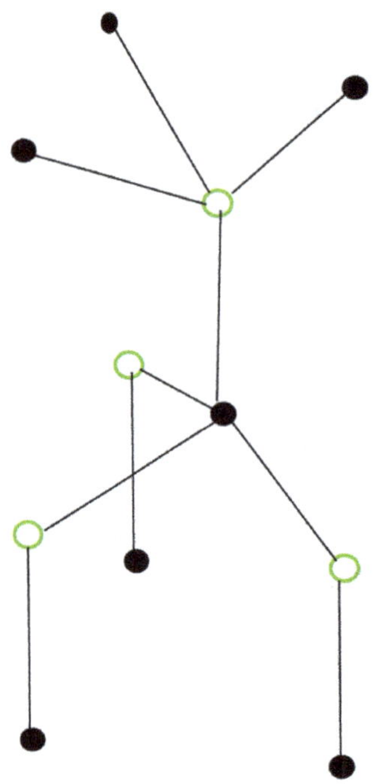

Figure 8. The unit cell of the wurtzite

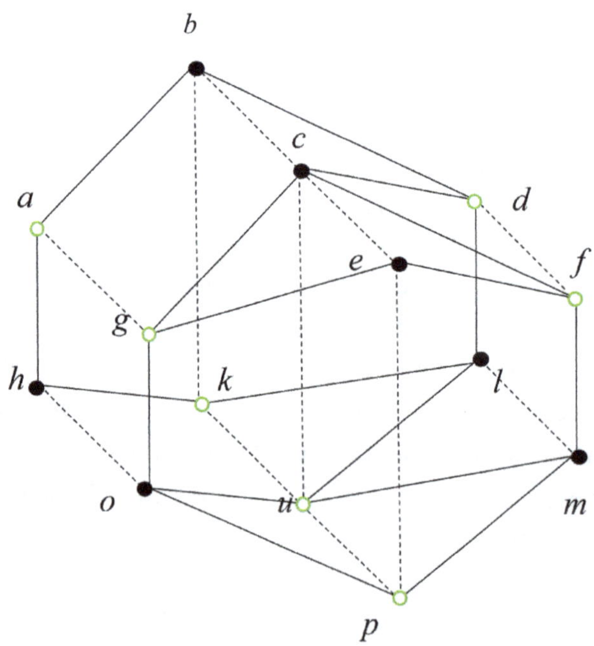

$14 - 29 + 21 - 6 = 0.$

Thus, the dimension of polytope on Figure 8 is equal to 4, i.e., the unit cell structure of the wurtzite has dimension 4.

The Dimension of the Fluorite

On example of compound $MnCl_2$ we look at the structure of fluorite. Isolate magnesium atoms lying at the centers of the cube faces ( ), and chlorine atoms (▲), forming a smaller cube inside the bigger cube, which are located at the vertices of magnesium atoms (Figure 9).

From Figure 9 it follows that the number of vertices is 14, i.e. $f_0 = 14$, the number of edges is 36, i.e. $f_1 = 36$, the number of flat faces is sum from number of triangles (24) and number of rectangles (6) (smaller cube faces), i.e. $f_2 = 30$. The number three dimension shape to sum up from smaller cube (1), pyramids on its faces (6) and figure (22) without inner parts (1), i.e. $f_3 = 8$

Figure 9. The unit cell of the fluorite
- magnesium atom, ▲ - chlorine atom

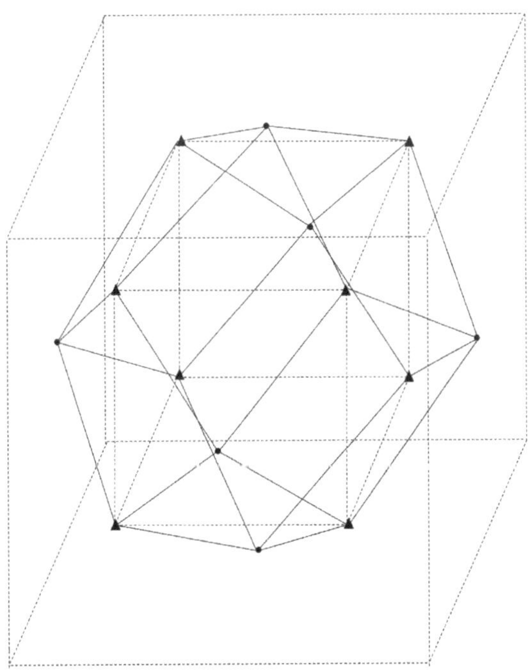

8. Substituting these values $f_i, (\mathrm{i} = 0,1,2,3)$ in the equation (2), we find that it is satisfied for $n = 4$

$$14 - 36 + 30 - 8 = 0.$$

Thus, the dimension of polytope on Figure 9 is equal to 4, i.e. the unit cell structure of the fluorite has dimension 4.

REFERENCES

Arcel, V. (1931). *Chemical Bond. Gannover* Higher Technical School.

Fersman, A. E. (1937). *Geochemistry* (Vol. 3). Leningrad: ONTI.

Gillespie, R. J. (1972). *Molecular Geometry*. New York: Van Nostrand Reinhold Company.

Gillespie, R. J., & Hargittai, I. (1991). *The VSEPR Model of Molecular Geometry*. London: Allyn & Bacon.

Gray, H. B. (1965). *Electrons and chemical bonding*. New York: W.A. Benjamin INC.

Hund, F. (1927). *Linienspektren und periodisches System der Elements*. Gottingen: Springer. doi:10.1007/978-3-7091-5695-7

Karapet`yants, M. X., & Drakin, S. I. (1994). *General and Inorganic Chemistry*. Moscow: Chemistry.

Mendeleev, D. I. (1934). *The periodic law of chemical elements*. Moscow: Gos. Chemical and Technical Publishing.

Metzler, D. E. (1980). Biochemistry. The Chemical Reactions of Living Cells (vols. 1-3). New York: Academic Press.

Pauli, W. (1925). Uber den Zusammen hang des Abschusses der Elektronengruppen in Atommit der Komplexstruktur der Spetren. *Zeitschrift fur Physik, 31*(1), 765–783. doi:10.1007/BF02980631

Perrin, D. D., Armarego, W. L. F., & Perrin, D. R. (1980). *Purification of Laboratory Chemical*. New York: Pergamon.

Poincare A. (1895). Analysis situs. *J. de é Ecole Polytechnique, 1,* 1 – 121.

Vernadsky, V. I. (2012). *The Biosphere*. London: Springer.

Zhizhin G.V. (1998). *Chemistry of d- and f-elements.* Sankt-Petersburg: Nord-West Polytechnic Institute.

Zhizhin, G. V. (2014a). On the higher dimension in nature. *Biosphere*, *6*(4), 313–318.

Zhizhin, G. V. (2014b). *World – 4D*. St. Petersburg: Polytechnic Service.

Zhizhin, G. V. (2015, November). *Polytopic prismahedrons – fundamental regions of the n-dimension nanostructures.* Paper presented at The International conference "Nanoscience in Chemistry, Physics, Biology and Mathematics", Cluj-Napoca, Romania.

Zhizhin, G. V. (2016). The structure, topological and functional dimension of biomolecules. *J. Chemoinformatics and Chemical Engineering*, *5*(2), 44–58.

Zhizhin, G. V., & Diudea, M. V. (2016). Space of Nanoworld. In M. V. Putz & M. C. Mirica (Eds.), *Sustainable Nanosystems, Development, Properties, and Applications* (pp. 214–236). New York: IGI Global.

Zhizhin, G. V., Khalaj, Z., & Diudea, M. V. (2016). Geometrical and topology dimensions of the diamond. In A. R. Ashrafi & M. V. Diudea (Eds.), *Distance, symmetry and topology in carbon nanomaterials* (pp. 167–188). New York: Springer. doi:10.1007/978-3-319-31584-3_12

KEY TERMS AND DEFINITIONS

Anomalous Elements: Transitional chemical elements in which the filling of the orbitals by electrons occurs with a violation of the experimental series of an increase in the energy of orbitals.

Established Rule Is the Filling by Electrons of the Orbitals of Anomalous *d*-Elements: in an anomalous *d*-elements is transition of one or two electrons from the *ns*-orbital on the $(n - 2)$ *d*-orbital with higher energy.

Established Rule Is the Filling by Electrons of the Orbitals of Anomalous *f*-Elements: In an anomalous *f*-elements is transition of one electron with the *nf*-orbital on the $(n + 1)$ *d*-orbital with higher energy.

Geometrical Image of a Chemical Compound: The geometrical image of a chemical compound (molecule) is a convex polytope, at the vertices of which atoms (or functional groups) are located. The edges of the polytope connecting the vertices correspond to the chemical bonds of the compound. The part of edges only carry a geometric function. They are necessary to give the molecule the image of a convex geometric figure. The dimension of the polytope is determined by the Euler-Poincare equation.

Magnetic Quantum Number m_l**:** Characterizes of the orbital orientation in space.

Orbital Quantum Number l**:** Characterizes of the form of the electron cloud (orbital or subshell).

Principal Quantum Number n**:** Characterizes of the energy level (shell) of electrons in atom.

Spin Quantum Number m_s**:** Characterizes of the spin orientation.

Transitional Elements: Chemical elements in which electrons fill *d*- and *f*-orbitals of an atom (*d*- and *f*-elements).

Chapter 2
The Structure and Higher Dimension of Molecules *s*- and *p*-Elements

ABSTRACT

There are considered chemical compounds in which s- and p-elements participate, i.e., elements in which electrons fill with the s- and p-orbitals of atoms. Many of these elements, showing increased chemical activity, play an important role in the vital activity of living organisms and are included in drugs for the treatment of living organisms. The structures of these compounds have been determined and classified, and the molecules of these compounds have been shown to have both rule of higher dimensionality (4, 5, 6, and more). This can be of significant importance for nanomedicine.

THE STRUCTURE AND HIGHER DIMENSION OF ALKALINE METALS

In the vast majority of compounds involving alkali metals (elements of first group of the Mendeleev table) the chemical bond is preferably ionic. Alkali metals have an external electron shell ns^1. They easily give bake one electron exhibiting a degree of oxidation +1. Salts of alkali metals in the condensed state usually have a cubic lattice, forming a structure of type rock salt structure (Chapter 1, Table 3).

DOI: 10.4018/978-1-5225-4108-0.ch002

However, Oganov and his co-workers have found that under high pressure the structure of many compounds, including alkali metal compounds, acquires new unexpected properties (Zhung et al., 2013; Zhou et al., 2012; Zhu, Oganov & Lyakhov, 2013). In particular, it is shown that the structure of the sodium-chlorine compounds varies significantly (Zhung at all, 2013). The elementary cell of this structure is a cube with sodium atoms at its vertices, and an icosahedron with a center with chlorine atoms at its vertices is located inside the cube. This compound is denoted Pm3 – $NaCl_x$.

Theorem 1 (Zhizhin, 2016a)

The dimension of the unit cell Pm3 – $NaCl_x$ is 5.

Proof

In the proof of Theorem 1 we shell use equation Euler- Poincare (2) in Chapter 1.

Figure 1 shows the structure of this compound, where sodium atoms are located at the vertices of the cube 13 -19, and chlorine atoms are located at the vertices 1 - 12, 21.

Note that the icosahedron with the center already has dimension 4. Indeed, the icosahedron has 12 vertices, 30 edges, 20 flat faces. If you enter a center in the icosahedron (point 21), then $f_0 = 13, f_1 = 42$, $f_2 = 50$ for it. In addition, 20 tetrahedrons are added, taking into account that they are located in the icosahedron, we obtain $f_3 = 21$. Substituting the obtained values of f_i into equation (2) of Chapter 1, we find that it is satisfied for $n = 4$

$$13 - 42 + 50 - 21 = 0.$$

This proves that the icosahedron with the center has dimension 4.

From it follows that the dimension of the polytope with 21 vertex in Figure 1 is greater than 4. To determine this dimension, let us calculate the number of elements of different dimension entering into this polytope. Thus, for this polytope $f_0 = 21$. The number of edges is the sum of the number of edges of the icosahedron (30), the number of edges issuing from the center to the vertices of the icosahedron (12), the number of edges of a cube (12), the number of edges issuing from the vertices of the cube to the vertices of the icosahedron (24). Hence, $f_1 = 78$.

Figure 1. The structure of compound sodium and chlorine at high pressure

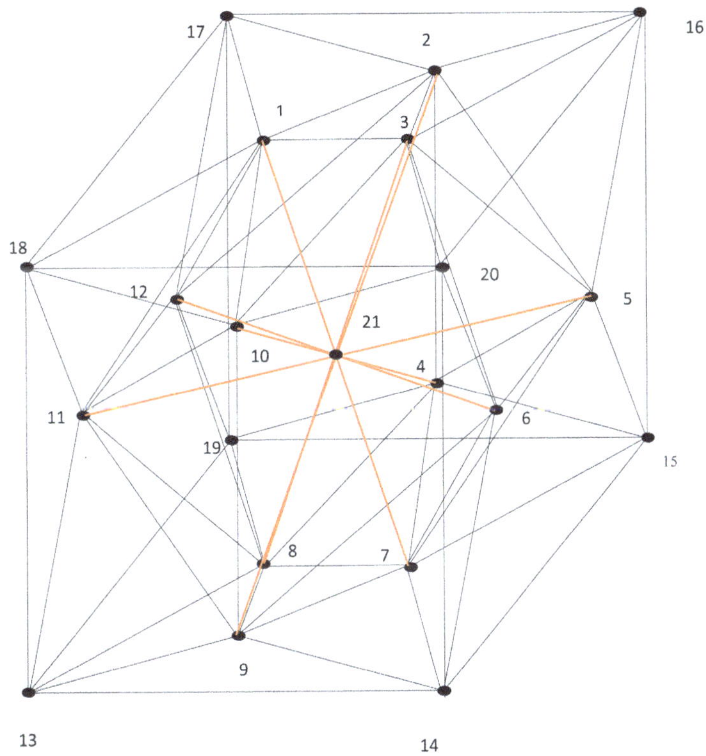

Two-dimensional elements include trapezoids: 13-8-7-14, 1-3-17-16, 1-3-18-20, 19-8-7-15, 11-12-18-17, 11-12- 19-13, 6-5-20-16, 5-6-14-15, 18-10-9-13, 9-10-20-14, 17-2-4-19, 2-4-15- 16. Total number of trapezoids is 12.

In the number of two-dimensional elements includes also triangles:

1. The triangles of the outer surface of the icosahedron (20);
2. The triangles inside the icosahedron ((20 • 3) / 2 = 30);
3. Triangles of tetrahedrons resting on the faces of icosahedrons (with the exception of the icosahedron faces themselves) having common vertices with a cube: 13-8-9, 13-11-9, 13-8-11, 14-7-9, 14-7 -6, 14-9-6, 15-4-5, 15-4-7, 15-5-7, 19-4-8, 19-12-8, 19-12-4, 17-1-12, 17-1-2, 17-12-2, 16-3-2, 16-2-5, 16-3-5, 18-1-11, 18-1-10, 18-10-11, 20 -3-10, 20-3-6, 20-6-10; all these triangles is 24;
4. Triangles of pyramids resting on the cube's faces 13-1-18, 17-12-19, 18-17-1, 20-3-16, 16-5-15, 20-6-14, 19-8 -13, 14-7-15, 18-10-20, 14-7-15, 17-2-16, 19-4-1; all of these triangles is 12.

5. The total number of triangles is 20 + 30 + 24 + 12 = 86.

Also in the number of two-dimensional elements includes 6 squares of cube faces. The total number of two-dimensional elements is f_2 = 12 + 86 + 6 = 104.

In the number of the three-dimensional elements includes:

- 1 cube,
- 1 icosahedron,
- 20 tetrahedrons in the icosahedron,
- 8 tetrahedrons at the tops of the cube 17-1-2-12, 16-2-3-5, 15-4-5-7, 14-6-7- 9, 13-11-8-9, 18-1-12-11, 19-12-4-8, 20-3-6-10;
- 6 pyramids with a base face of the cube 13-11-18-19-12-17, 19-17-4-2-15-16, 16-15-5-6-14-20,13-18-10-9- 20-14, 18-17-1-3-20-16, 13-19-8-7-14-15;
- 12 pyramids on the trapezes of these pyramids 18-17-11-12-1, 13-18-11-10-9, 11-12-13-19-8, 20-14-10-9-6, 17-16- 1-3-2, 18-20-1-3-10, 13-14-9-8-7, 19-15-8-7-4, 14-15-7-6-5, 20-16- 6-5-3, 16-15-5-2-4, 19-17-12-2-4.

The total number of three-dimensional elements is f_3 = 2 + 20 + 8 + 6 + 12 = 48.

The four-dimensional element, as already proved, is an icosahedron with a center. There are other four-dimensional elements. The second such element is the polytope in Figure 1 after removing the center. Indeed, in this case f_0 = 21-1 = 20, f_1 = 78-12 = 66, f_2 = 104-30 = 74, f_3 = 48-20 = 28. Substituting these values of f_i into equation (2) of Chapter 1, we find that it is satisfied for $n = 4$

$$20 - 66 + 74 - 28 = 0.$$

This proves that the figure in Figure 1 after removing the center is a polytope of dimension 4. We can find three more elements of dimension 4 in Figure 1 if we separate the prism connections on the parallel faces of the cube: upper and lower, right and left, front and back. Since these constructions are compatible with a rotation by 90^0, we prove the desired equality for only one of these constructions, for example, for the upper and lower faces of the cube. This construction is shown in Figure 2.

Figure 2. Four-dimensional part of the unit cell of the compound sodium and chlorine at high pressure

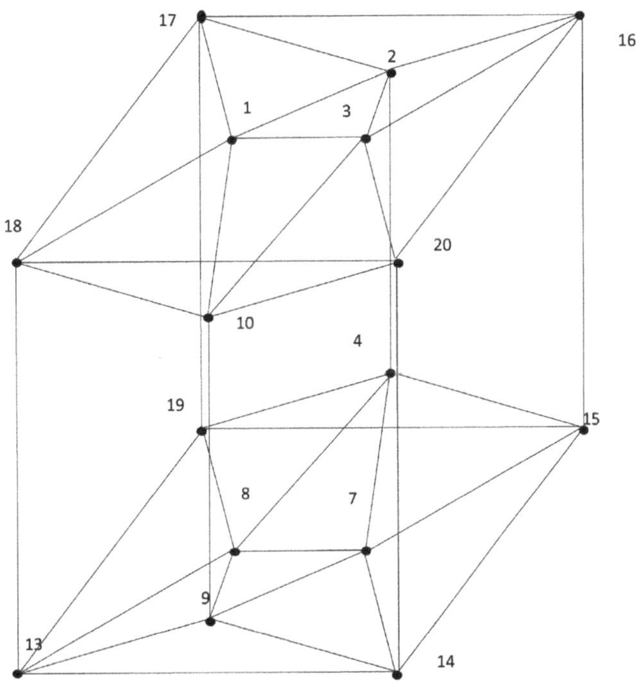

In it the number of vertices is 16, i. e. $f_0 = 16$. The number of edges consists of the number of edges of the cube (12); the number of edges of the cube in the upper part (without the edges of the cube) is 17-2, 2-16, 17-1, 1-2, 2-3, 1-3, 3-20, 3-10. 1-10, 1-18, 18-10, 10-20, 3-16, i.e. 13 edges; the same number of edges (13) in the lower part; 2 connecting (vertical) edges 2-4, 10-9. Thus, the total number of edges is $f_1 = 40$.

In the number of two-dimensional elements includes:

- 6 faces of the cube;
- 10 facets of the pyramids in the upper part of the structure and 2 triangles from the pyramid with a crouching upper bound of the cube, i.e. 12 two-dimensional elements;
- 12 two-dimensional elements in the lower part of the structure;
- 2 vertical trapezes at the back wall of the cube and 2 vertical trapezes near the front wall of the cube.

Thus, the total number of two-dimensional elements is $f_2 = 34$.
In the number of the three-dimensional elements in Figure 2 includes:

- 2 pyramids with a trapezoidal base and one pyramid with a cube face at the top of the structure;
- 3 the same pyramids in the lower part of the structure;
- 2 pyramids with the bases of the back and front walls of the cube;
- 1 cube;
- the figure left from the cube after deducting all the pyramids from it.

Thus, the total number of three-dimensional elements $f_3 = 10$. Substituting the values of f_i, determined for the polytope in Figure 2, into equation (2) of Chapter 1, we find that it is satisfied for $n = 4$

$$16 - 40 + 34 - 10 = 0.$$

This proves that the construction in Figure 2 has dimension 4. Taking into account the existence of two more similar constructions and the impossibility of the existence of other similar constructions, we conclude that for the polytope in Figure 1 $f_4 = 5$. Substituting the values of f_i, determined for the polytope in Figure 1, into equation (2) of Chapter 1, we find that it is satisfied for $n = 5$

$$21 - 78 + 104 - 48 + 5 = 2.$$

This proves that the polytope in Figure 1 has dimension 5 and it proves the theorem 1.

It should be expected that other alkali metal salts at high pressure have structures with elementary cells of higher dimension. Besides atoms of alkali metals enters into complex compounds with many of different elements. We shall see later that these compounds also have higher dimensionality at normal pressures.

THE STRUCTURE AND HIGHER DIMENSION OF COMPOUNDS ELEMENTS OF SECOND GROUP OF THE MENDELEEV TABLE

The elements of the second group of the periodic table have on the outer shell two of the valence electrons and can form linear molecules. For example, beryllium forms linear molecules with halogens BeF_2, $BeCl_2$, $BeBr_2$, BeI_2. However, in the vast majority of its compounds, beryllium manifests around itself tetrahedral coordination. An important example of such a compound is beryllium oxide. This compound is a valuable optical material. It is used in nuclear power engineering, microelectronics, and laser technology. The structure of this compound is a wurtzite structure. It shows tetrahedral coordination both around the beryllium atom and around the oxygen atom. The dimension of the wurtzite molecule was determined in Chapter 1, Figure 8. It turned out to be equal to 4. Thus, the dimension of the beryllium oxide molecule is 4. For beryllium oxide, atoms of beryllium are located at the vertices *b, c, e, h, l, m, o* of the polytope in Chapter 1, Figure 8, and oxygen atoms are located at the vertices *a, d, g, k, p, f*. The nature of the chemical bond in beryllium oxide is still open. Estimates of the relative contribution of the ionic and covalent bonds are contradictory (Sholl & Walter, 1969; Hidaka, 1976). In any case, it is impossible to explain the observed structure of beryllium oxide by any distribution of electrons, divided or unshared electron pairs over elementary quantum cells of the adopted system of electronic orbitals. However, this also applies to compounds of the wurtzite type, discussed in Chapter 1.

When the beryllium oxide is an extended crystalline body, then in order to build a model of such a body, the polytope in Chapter 1, Figure 8 need to multiply by an edge and to obtain a polytopic prismedron of dimension 5. This prismahedron is a stereohedron and it translation it is possible to fill the space without gaps (see Chapter 7).

In crystalline beryllium fluoride, the linearity of the combination of beryllium atoms with fluorine atoms and the tetrahedral coordination of beryllium atoms with one another surprisingly are combined. In this case, a structure is formed topologically equivalent to the adamantane molecule, which includes 10 carbon atoms (see Figure 15). The difference from the adamantane molecule is the linear arrangement between the beryllium atoms of the fluorine atoms (Figure 3). Since the adamantane molecule has a dimensionality of 4 (Zhizhin, 2014a), the unit cell dimension in crystalline beryllium fluoride is also 4.

The second element in the second group of the periodic table after beryllium is the element of magnesium. Like beryllium, in accordance with the arrangement in the second group, it has two valence *s* - electrons on the outer layer. However, in contrast to beryllium, it has a completely filled pre-extrinsic layer of electrons. This layer includes two *s* - electrons and six *p* -electrons. These four electron pairs, starting from each other, create

Figure 3. The structure of crystalline beryllium fluoride

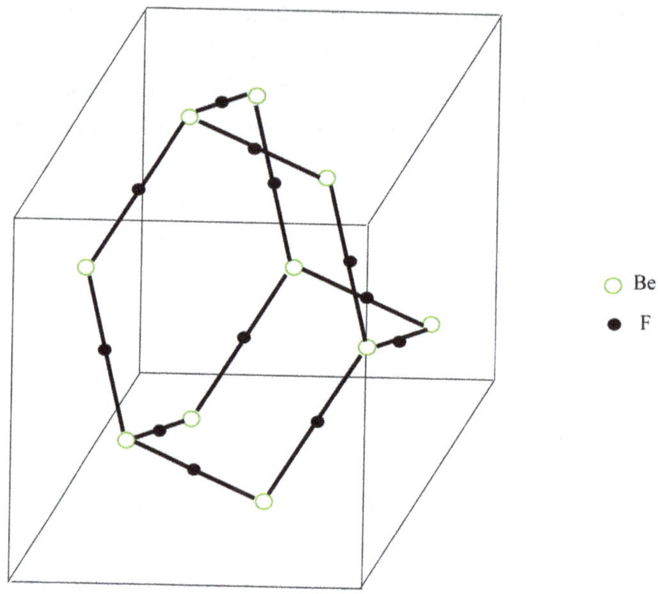

tetrahedral coordination around the magnesium atom. Taking vacant quantum cells of ligands, they increase the possible value of valence of magnesium to six. This gives magnesium more chemical activity especially important for living organisms. It participates in all metabolic processes in living organisms. Magnesium is one of the basic elements of the cell. It stimulates the work of enzymes that break down proteins and other nutrients. Magnesium participates in the harmonious work of all body systems, especially the central and peripheral nervous system, affects the growth of estrogen hormones and blood coagulability.

Even for the chemical bonds of magnesium with valence 2, compounds of higher dimension are formed. Consider a molecule of bis (neopentyl) magnesium $Mg(C_5H_{11})_2$ (Gillespie & Hargittai,, 1991). Magnesium in this molecule exhibits a valence of 2. In each group C_5H_{11}, the carbon atoms form the geometric form of a tetrahedron centered. This already gives the dimension of this form equal to 4. In addition, around each carbon atom there is also a tetrahedral coordination of other atoms (hydrogen and carbon). Each group C_5H_{11} can be represented in the form of a tetrahedron with a center in which its vertices contain functional groups CH_3, and in the fourth (attached to the magnesium atom) is a functional group CH_2. At the center of the tetrahedron is a carbon atom. Then the bis (neopentyl) magnesium molecule has the form of two tetrahedrons with a center connected to each other by a magnesium atom (Figure 4). Functional groups CH_3 are located in the vertices $a_1, c_1, d_1, a_2, c_2, d_2$; functional groups CH_2 are located in the vertices b_1, b_2; at the points o_1, o_2 are carbon atoms; at the point o there is a magnesium atom.

Valentine bonds are indicated in Figure 4 with a brown color. The remaining edges (black) serve to form a convex figure (polytope), the dimension of which must be established.

Theorem 2

The dimension of bis(neopentyl) magnesium molecule equal to 6.

Proof

For proof of theorem 2 we noted that polytope on Figure 4 is 5-cross-polytope with centrum (Figure 5).

Comparing Figures 4 and 5, we see that these figures are topologically equivalent, that is, in Figure 5, the same vertices are shown as in Figure 4. Moreover, each of the corresponding vertices in Figure 5 is incidental to the number of edges as in Figure 4 and the connection of vertices by edges in

Figure 4. The structure of bis(neopentyl) magnesium molecule

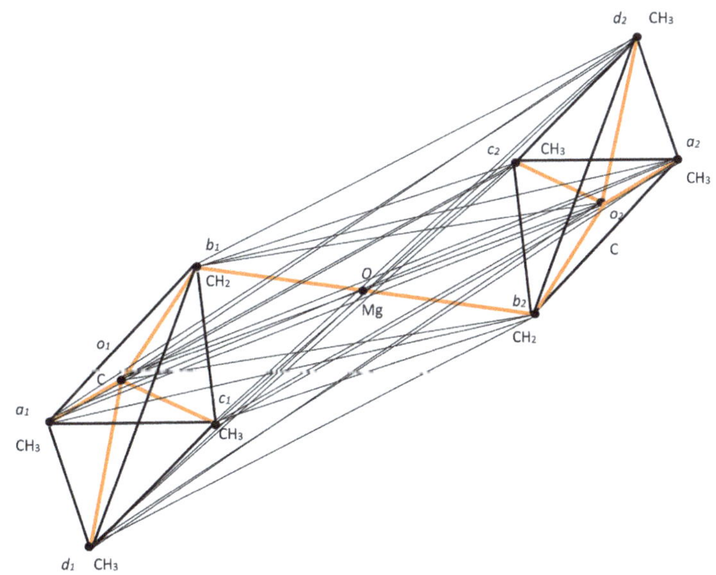

Figure 5. The 5 –cross-polytope with centrum

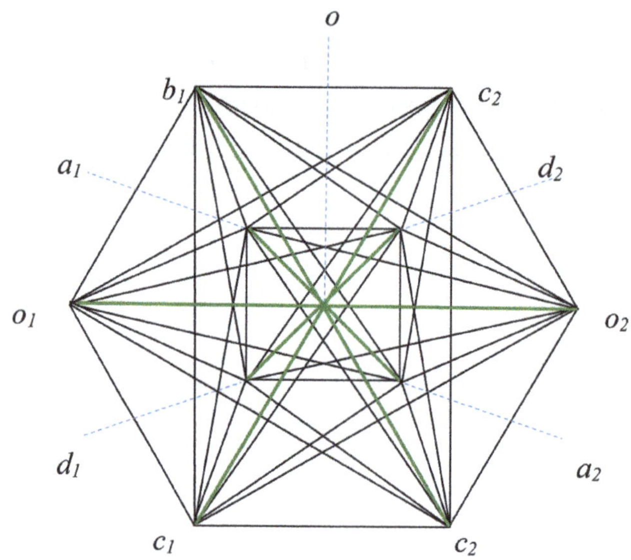

Figure 5 is topologically the same as in Figure 4. If we denote in Figure 5 the edges issuing from vertex O to other vertices in green, the remaining figure, as can be seen, is the 5-cross - polytope, given in the monograph by Zhizhin, (2014b). In addition, the vertex O is the center of 5 - cross-polytope. As follows from Zhizhin (2014b) 5 cross-polytope has 10 vertices $(f_0 = 10)$, 40 edges ($f_1 = 40$), 80 triangular faces ($f_2 = 80$), 80 tetrahedrons ($f_3 = 80$), 32 4-cross-politopes ($f_4 = 32$). The introduction of the center into the 5-cross-polytop adds, according to Figure 5, 10 edges

$(oa_1, ob_1, oc_1, od_1, oo_1, ob_2, oa_2, oc_2, od_2, oo_2)$,

24 triangular faces

$\{$
$o_1b_1o, b_1a_1o, b_1oa_2, b_1od_2, b_1oc_2, b_1oc_1, c_2d_2o, c_2od_1, c_2oa_2, c_2oo_2, o_2d_2o, o_2oa_2, o_2ob_2, b_2a_2o, b_2od_2,$
$b_2oc_2, b_2oc_1, b_2d_1o, c_1od_1, c_1od_2, c_1oa_2, c_1oo_1, a_1o_1o, o_1d_1o$ $\}$,

28 tetrahedrons

$(b_1od_2a_2, b_1c_1oo_1, b_1d_1oa_1, b_1a_1d_2o, b_1od_1a_2, b_1oa_2c_2, b_1od_1c_1, c_2oa_1d_1, c_2b_2oo_2,$

$c_2a_2od_2, c_2a_1d_2o, c_2oa_2d_1, c_2od_1b_1, c_2oa_2b_2, c_1oa_2d_2, c_1a_1od_1, c_1d_1a_2o, c_1oa_1d_2, c_1od_2b_2, c_1od_1b_1,$

$o_1a_1d_1o, b_2od_1a_1, b_2d_2oa_2, b_2a_2d_1o, b_2od_2a_1, b_2oa_1c_1, b_2od_2c_2, o_2d_2a_2o$),

18 4-siplexes

($b_1a_1od_2c_2, b_1c_1d_1a_1o, b_1o_1a_1oc_2, b_1od_2o_2c_2, b_1od_2a_2o_2, c_1d_1oa_2b_2, c_1o_1d_1b_2o,$

$c_1oa_2o_2b_2, c_1oo_2d_2o_2, c_2b_2a_2d_2o, c_2o_2d_2ob_1, c_2oa_1o_1b_1, c_2oa_1d_1o_1, b_2a_2od_1c_1, b_2od_1o_1c_1,$

$b_2od_1a_1o_1, o_2c_2od_2a_2b_2$),

6 5-siplexes

($o_1b_1a_1od_1c_1, c_2od_2a_2o_2b_2, c_1od_2a_2o_2b_2, b_2oa_1o_1d_1c_1, c_2od_1a_1o_1b_1, b_1oa_2o_2d_2c_2$).

Adding the obtained quantities of geometric figures of different dimensions connected with the center of the 5-cross polytope to the corresponding numbers of figures not connected with the center of the 5-cross polytope, we obtain the total number of geometric figures of different dimensions in the 5-crosspolytopewithcenter: $f_0 = 11, f_1 = 50, f_2 = 104, f_3 = 108, f_4 = 50, f_5 = 7$. Substituting these values into equation (2) of Chapter 1, we find that the Euler-Poincare equation is satisfied for $n = 6$

$$11 - 50 + 104 - 108 + 50 - 7 = 0.$$

This proves that a 5-cross-polytope with center has dimension 6. Theorem 2 is proved.

It should be noted that the above evidence accurately lists (in view of the work Zhizhin, 2014b) all the 108 three-dimensional figures included in the 6-dimensional 5-cross polytope with the center. This is significantly different from the proof of the existence of 4-dimensional 100-cell and 600-cell cells in Coxeter's work (Coxeter, 1963), for which the direct enumeration of three-dimensional figures included in these polytopes has not yet been received. Therefore, the question of proving the existence of these 4-dimensional polytopes remains open.

If the electron pairs of magnesium at the second energy level enter into a chemical bond, then its valence is more than two. For example, in Grignard reagent the magnesium valence is 4 and in the vicinity of magnesium atom there is tetrahedral coordination. While the nearest neighborhood of the magnesium atom has a dimension of 4, and with the account of the attached groups of atoms this dimension is even higher. An interesting example is the complex magnesium ion $Mg(OAsMe_3)_5^{2+}$, Me = CH_3. In this compound, magnesium exhibits a valence of 5. In this case, the nearest environment of magnesium is of dimension 5. Indeed, the nearest environment of magnesium by oxygen atoms has the form of a 4-simplex with a center in the magnesium atom (Figure 6). At the vertices *a, b, c, d, e* of the polytope, in Figure 6, there are oxygen atoms, in the vertex *o* there is a magnesium atom. The valence bonds are indicated in Figure 6 with a brown color, the other edges (black) are needed to create a convex figure in space. The vertices together with the connecting ribs form a 4-simplex. The addition of a magnesium atom and valence bonds converts this polytope into a 4-simplex with a center.

In Figure 6 can to indicate 6 vertices (f_0 = 6); 15 edges (f_1 = 15);

20 trigonal faces (*abc, aeb, abo, abd, bcd, bco, bce, aeo, aed, aec, edo, edc, edb, dco, dca*), $f_2 = 20$;

15 tetrahedrons (*abed, abec, abcd, dbce, aecd, obcd, oecd, aoed, aoeb, aobc, boed, coae, doeb, eobc, aocd*), f_3 = 15;

Figure 6. The 4 –simplex with centrum

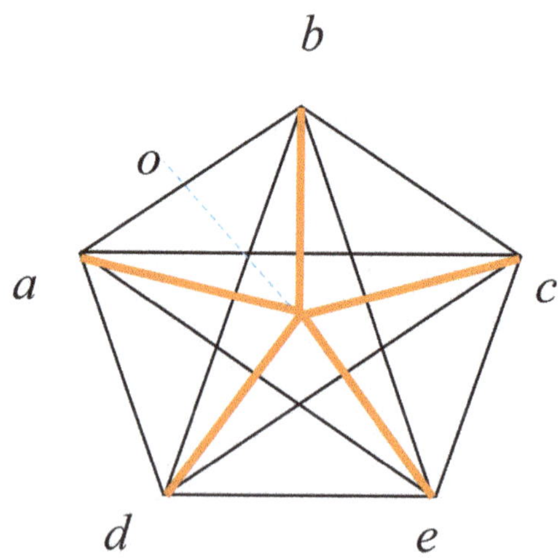

6 4-simplexes (*abcde, abedo, abeco, abcdo,dbceo, aecdo*), $f_4 = 6$.

Substituting values f_i into equation (2) of Chapter 1, we find that the Euler-Poincare equation is satisfied for $n = 5$

$$6 - 15 + 20 - 15 + 6 = 2.$$

This proves that a 4-simplex with center has dimension 5. If we take into account the presence of other atoms in the ion $Mg(OAsMe_3)_5^{2+}$, then its dimension will be even higher.

Such compounds can form other alkaline-earth elements, i. e. calcium and barium.

THE STRUCTURE AND HIGHER DIMENSION OF COMPOUND ELEMENTS OF THE GROUP THREE (a) OF THE MENDELEEV TABLE

The first element of group 3a of the element table is boron. Like all elements of this group, it has two *s*-electrons and one *p*-electron on the outer layer. Boron does not have vacant *d*-and *f*-orbitals, and there are not several electron pairs on the pre-existing layer, as, for example, for the atoms of alkaline-earth elements. However, the property of the collective interaction of electron pairs is also manifested here, but in a slightly different way compared to magnesium. Here pairs of electrons of the second energy level of several boron atoms interact, creating (repelling from each other) tetrahedral coordination of boron atoms. Therefore, in the compound B_4Cl_4, the boron atom has an effective valence of 4, and not three, which would correspond to the group number (Figure 7).

At the vertices *g, h, f, e* in Figure 7 boron atoms are located, and at the vertices *a, b, c, d* chlorine atoms arranged.

Theorem 3

The B_4Cl_4 molecule has dimension 4.

Figure 7. The structure of the B_4Cl_4 molecule

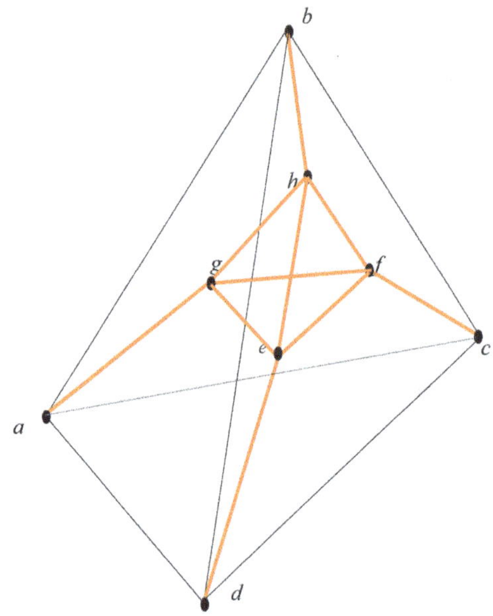

Proof

In Figure 7 there is eight vertices, $f_0 = 8$. The number of edges is 16 (*ab, bc, cd, ad, bd, ac, gh, hf,ef, he, gf, eg, bh, fc, ed, ag), $f_1 = 16$*. The number of elements of dimension 2 is 14 (triangles *abd, bcd, abc, acd, ghe, hef, ghf, gfe* and quadrangles *aghb,aged, hbed, hbfc, efcd, hfbc*), $f_2 = 14$. The number of elements of dimension 3 is 6 (tetrahedrons *abcd, ghef* and prismatoides *abhged, hbedbfc, aghbfc, gefadc*), $f_3 = 6$. On Figure 7 the edges correspondent of chemical bounds is indicated brown, remain edges (black) it is need for creating convex body. Substituting the values of the number of elements of different dimension in the equation Euler –Poincare (2) of Chapter 1, we obtain

$$8 - 16 + 14 - 6 = 0.$$

We find that it holds for $n = 4$. This proves that the figure who projection is shown in Figure 7 there is polytope of dimension 4. This proves theorem 3.

Due to the interaction of electron pairs of several atoms, formation of other compounds is also possible. For example, in Figure 8. The image of the B_6Cl_6 molecule is shown. Here, also, the edges corresponding to the chemical bonds is indicated in brown, the remaining edges are necessary for obtaining a convex figure.

In the compound, both the boron atoms and the chlorine atoms have octahedral coordination. The effective valence of boron in this compound is 5. In the polytope in Figure 8, boron atoms are located at the vertices $a_1, b_1, c_1, d_1, e_1, f_1$ and hydrogen atoms are located at the vertices $a_2, b_2, c_2, d_2, e_2, f_2$.

Theorem 4

The B_6Cl_6 molecule has dimension 4.

Proof

In this case the number of elements of zero dimension is $f_0 = 12$. The number of elements of dimension one is $f_1 = 12 + 12 + 6 = 30$. The number of elements of dimension 2 is sum of the number small triangles 8 and big triangles 8,

Figure 8. The structure of the B_6Cl_6 molecule

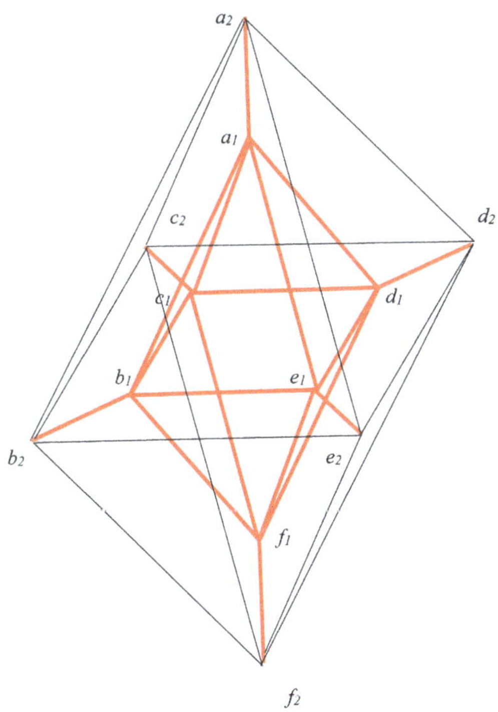

add 12 quadrangles, i. e. $f_2 = 28$. The number of elements of dimension 3 is sum two octahedrons and 8 prism, i.e. $f_3 = 10$. Substituting the values of numbers of elements of different dimensions in the equation (2) of Chapter 1, we obtain

$$12 - 30 + 28 - 10 = 0.$$

We find that it holds for $n = 4$. This proves that the figure 8 is polytope of dimension 4. This proves theorem 4.

Elements Al, Ga, In and Tl have vacant d - and f - orbitals and tend to supplement their valence shell to 6 electron pairs, and in several compounds In and Tl have more than 6 electron pairs. These elements in many compounds exhibit tetrahedral coordination in the vicinity of the atom. Taking into account the possible addition of other elements to tetrahedral coordination, complex compounds with high dimensionality can arise. For example, aluminum (a biogenic element) forms a cyclic compound $[(CH_3)_2AlF]_4$ (Figure 9).

If we form a convex figure from Figure 9, we get the polytope shown in Figure 10. At the vertices of a_1, a_4, a_7, a_{10} fluorine atoms are located. A · t the vertices $a_{13}, a_{14}, a_{15}, a_{16}$ aluminum atoms are located. Functional groups CH_3 are located in the $a_2, a_3, a_5, a_6, a_8, a_9, a_{11}, a_{12}$ vertices.

Theorem 5

The polytope of cyclic compound $[(CH_3)_2AlF]_4$ has dimension 5.

Figure 9. A cyclic compound $[(CH_3)_2AlF]_4$.

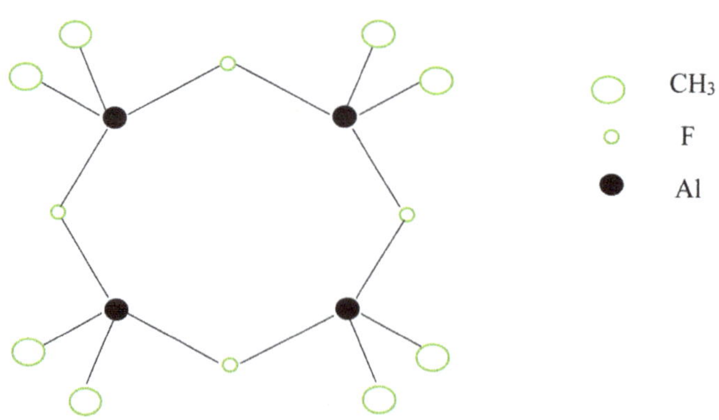

Figure 10. The convex polytope of cyclic compound $[(CH_3)_2AlF]_4$.

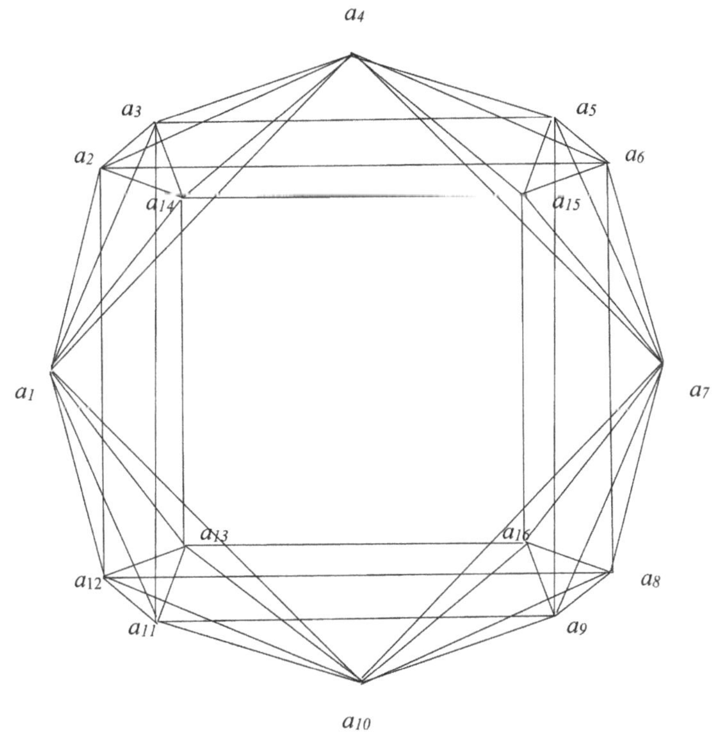

Proof

The polytope in Figure 10 has 16 vertices, $f_0 = 16$; 52 edges, $f_1 = 52$. In addition, it has 4 polytopes of dimension 4 each (tetrahedrons with a center) $a_1a_2a_3a_4a_{14}, a_4a_5a_6a_7a_{15}, a_7a_8a_9a_{10}a_{16}, a_{10}a_{11}a_{12}a_{13}$. Each tetrahedron with a center has 10 triangular faces. This gives 40 triangular faces in the polytope 10. In addition, three triangular faces are formed at the vertices a_1, a_4, a_7, a_{10} with horizontal and vertical sides. This gives another 4 3 = 12 triangles. There are 4 more rectangular faces

$(a_{13}a_{14}a_{15}a_{16}, a_6a_2a_8a_{12}, a_{11}a_5a_3a_9, a_1a_7a_{10}a_4)$

and 12 trapezoids

$(a_3a_5a_{15}a_{14}, a_3a_2a_5a_6, a_{15}a_2a_6a_{14}, a_6a_8a_{15}a_{16}, a_6a_8a_5a_9, a_{15}a_5a_9a_{16})$.

Thus, the total number of two-dimensional faces 68, $f_2 = 68$. Each tetrahedron with a center has 5 tetrahedrons. Therefore, the total number of tetrahedrons in Figure 10 is 5 • 4 = 20. Each of the vertices a_3, a_7, a_4, a_{10} is the vertex of the three pyramids. The total number of these pyramids is 12:

$$a_5a_{15}a_3a_4a_{14}, a_6a_2a_3a_4a_5, a_4a_2a_6a_{15}a_{14}, a_1a_2a_3a_{12}a_{11}, a_1a_{14}a_3a_{11}a_{13},$$

$$a_{10}a_{12}a_{11}a_8a_9, a_{10}a_{12}a_{13}a_8a_{16}, a_{10}a_{11}a_{13}a_{16}a_9, a_7a_8a_9a_5a_6, a_7a_9a_{16}a_5a_{15}, a_7a_8a_{16}a_6a_{15}.$$

There are four triangular prisms:

$$a_2a_3a_{14}a_5a_6a_{15}, a_{15}a_5a_6a_8a_9a_{16}, a_{11}a_{12}a_{13}a_8a_9a_{16}, a_2a_3a_{14}a_{11}a_{12}a_{13},$$

and six quadrangular prisms:

$$a_{13}a_{14}a_{15}a_{16}a_3a_5a_9a_{11}, a_{13}a_{14}a_{15}a_{16}a_2a_6a_8a_{12}, a_{13}a_{14}a_{15}a_{16}a_1a_4a_7a_{10},$$

$$a_2a_6a_8a_{12}a_3a_5a_9a_{11}, a_1a_2a_4a_6a_7a_8a_{10}a_{12}, a_1a_3a_4a_5a_7a_9a_{10}a_{11}.$$

Than the total number of three-dimensional figures is 42, $f_3 = 42$.

In addition to the 4 tetrahedrons mentioned with the center, as four-dimensional figures, there are another four-dimensional figures. In particular, this is a figure (F), shown in Figure 11. Indeed, this figure has 12 vertices, (F) = 12; 24 edges,(F) = 24; 19 two-dimensional faces,(F) = 19; and 7 three-dimensional figures, (F) = 7. Substituting these values into the Euler - Poincaré equation (2) of Chapter 1, we obtain that it is satisfied for $n = 4$

$$12 - 24 + 19 - 7 = 0.$$

This proofs that polytope F has dimension 4.

Four identical polytopes of dimension 4 exist in a neighborhood of each of the vertices a_1, a_4, a_7, a_{10}. One of these polytopes (L) is depicted in Figure 12. It has 12 vertices, f_0 (F) = 12; 24 edges, f_1 (F) = 24; 19 two-dimensional faces, f_2 (F) = 19; and 7 three-dimensional figures, f_3 (F) = 7. Substituting these values into the Euler-Poincare equation (2) of Chapter 1, we obtain

Figure 11. The 4 – dimension polytope F included in Figure 10

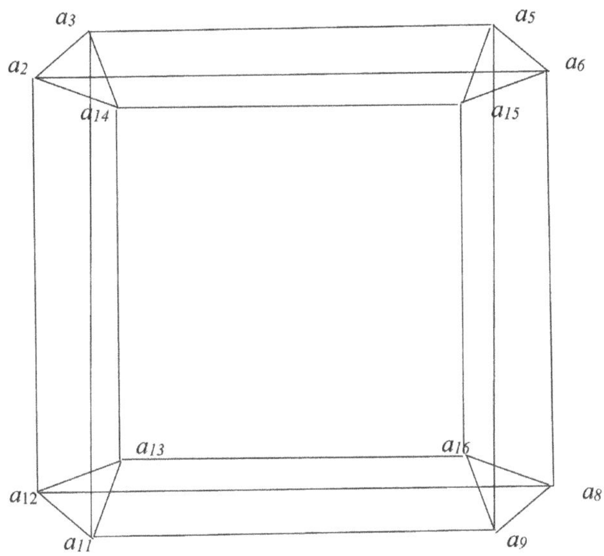

$$7 - 15 + 4 - 6 = 0,$$

i.e. the equation (2) hold for $n = 4$ and all the polytopes L has dimension 4.

Three more topologically equivalent polytopes of dimension 4 can be distinguished from Figure 10. Each of these polytopes consists of a rectangular prism and four tetrahedrons connected to each other in a cycle along the vertices of . These are polytopes

. One of them (polytope K) is shown in Figure 13.

The K polytope has 12 vertices, $f_0(K) = 12$; 32 edges, $f_1(K) = 32$; 31 two-dimensional faces, $f_2(K) = 31$; and 11 $3D$ facets, $f_3(K) = 11$. Substituting these values into the Euler-Poincare equation (2) of Chapter 1, we obtain

$$12 - 32 + 31 - 11 = 0,$$

i. e. the equation (2) of Chapter 1 hold for $n = 4$ and all the polytopes K has dimension 4.

Thus, the polytope on Figure 10 has 11 polytopes of dimension 4. Therefore, for polytope on Figure 10 the Euler-Poincare equation (2) has face

$$16 - 52 + 68 - 42 + 11 = 2,$$

Figure 12. The 4 – dimension polytope L included in Figure 10

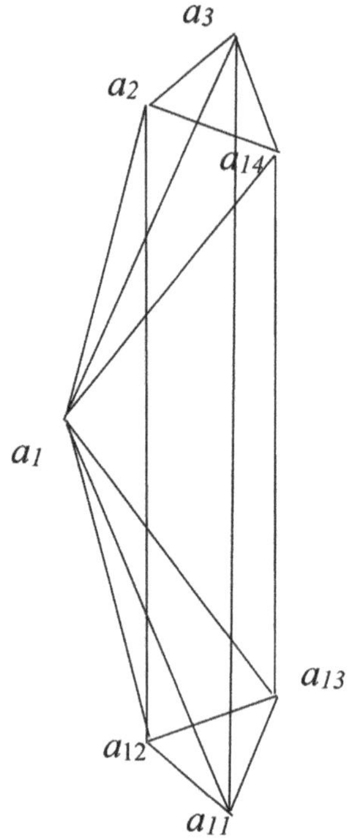

i.e. it hold for $n = 5$. This proofs theorem 5.

THE STRUCTURE AND HIGHER DIMENSION OF COMPOUNDS ELEMENTS OF THE GROUPS 4(a) – 7(a) OF THE MENDELEEV TABLE

The first element of group 4a of the Mendeleev table is carbon. This is the most important element on Earth. Any living organism consists largely of carbon. Carbon is the basis of all organic substances. In nature, carbon is found in the form of various minerals, the most important of which is diamond and graphite. Carbon is actively involved in chemical reactions, forming various compounds. There are many different allotropic forms of carbon. Recently, carbon is of considerable interest as the basis of nanomaterials (Zhizhin & Diudea, 2016b; Zhizhin, Khalaj & Diudea, 2016c). One of the molecular forms of carbon forming nanomaterials is the adamantane molecule. As a chemical compound, adamantine was discovered in 1933 (Landa & Machacek, 1933).

Figure 13. The 4 – dimension polytope K included in Figure 10

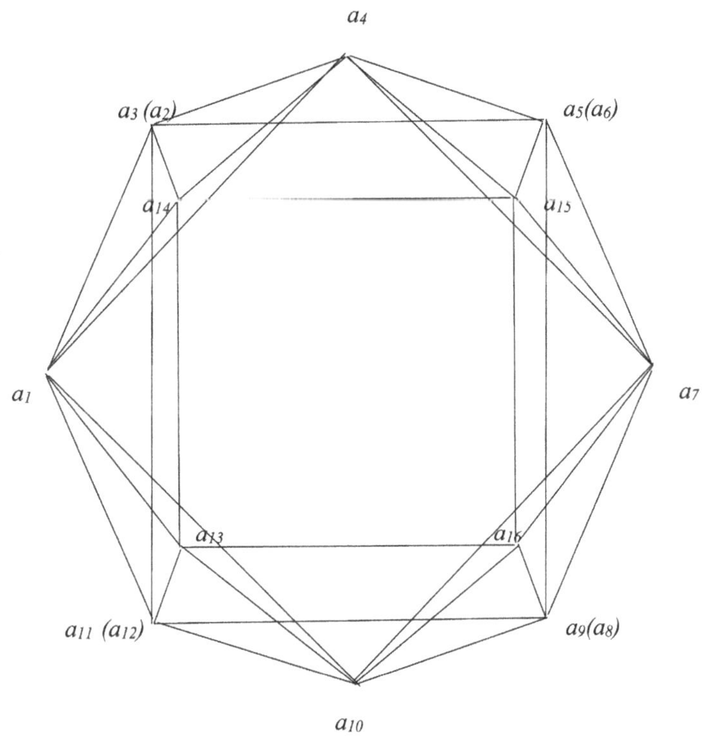

Adamantane molecule consists of 10 carbon atoms, repeating disposition of carbon atoms in the diamond crystal lattice, and 16 hydrogen atoms connected to carbon atoms by their valence links unsaturated with carbon atoms. Adamantane discovery served as an impulse for the development of organic polyhedranes chemistry. Derivatives of adamantane (e.g. amantadine, memantine, rimantadine, tromantadine) have found practical application in medicine as pharmaceuticals of different biological activity and purpose (as antiviral, antispasmodic, anti-Parkinson drugs, etc.). All these drugs have the same structural group of carbon atoms, which is peculiar to adamantane, only structural groups connected to carbon atoms change. Among inorganic and organoelemental compounds, there are many structural analogs of adamantane molecule, such as phosphorus oxide, urotropine and others. In 2005, a silicon analogue of adamantane has been synthesized (Fischer et al. 2005). In scientific literature (see Bauschlicher et al. 2007; Dahl et al. 2003) adamantane is usually depicted as it is shown in Figure 14.

Figure 14. Schema of the adamantane molecule

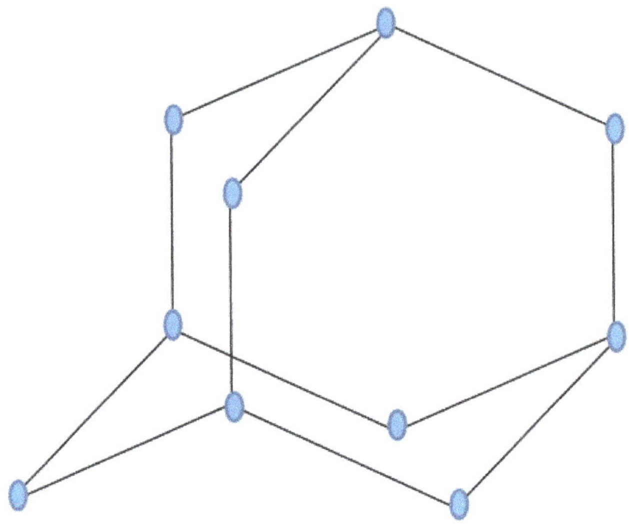

The adamantane structure is a common one. As a rule, hydrogen atoms are not depicted. Speaking hereinafter about adamantane molecule, we'll often keep in mind exactly 10 carbon atoms of adamantane molecule, although, strictly speaking, it is only a part of it. However, Figure 14 gives us little information and does not reflect the main features of spatial arrangement of atoms.

Theorem 6 (Zhizhin, 2014a)

The adamantane molecule is a convex polytope in the 4*D* space.

Proof

Let's construct an adamantane cell taking into consideration that 6 from 10 carbon atoms of adamantane are located in the centers of flat faces of the cube $\alpha_2, \alpha_6, \alpha_3, \alpha_4, \alpha_8, \alpha_9$. Each of the remaining four carbon atoms inside the cube $\alpha_1, \alpha_5, \alpha_7, \alpha_{10}$ is equidistant from the three nearest centers of the cube flat faces and relevant to the cube $\beta_1, ..., \beta_8$ vertices common to these faces (Figure 15).

On Figure 15 except for the edges of the cube (thin dashed line) are emerging in the construction of cell molecules lines of different kinds. Solid

Figure 15. The structure of molecule adamantane

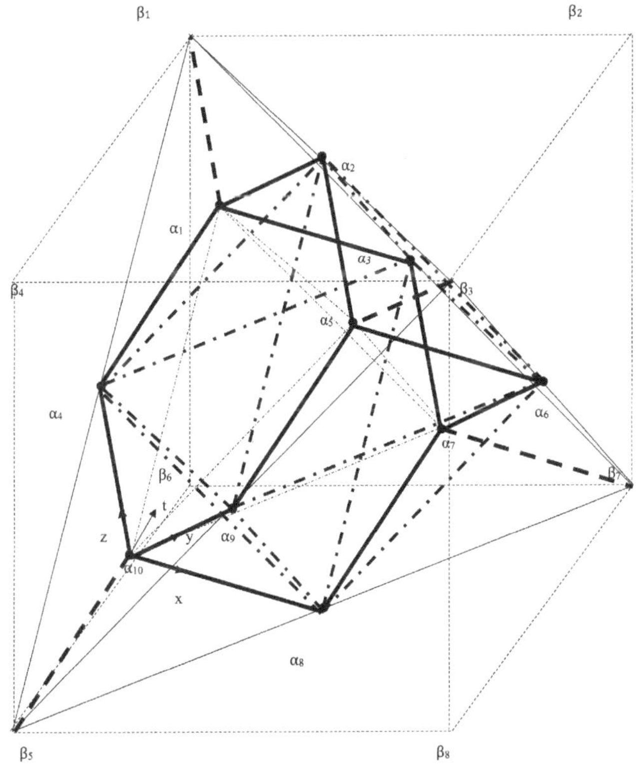

thick lines represent valence bonds between the carbon atoms; at three such links from each $\alpha_1, \alpha_5, \alpha_7, \alpha_{10}$ located inside $\beta_1, ..., \beta_8$ cube. The fourth connection from each of these carbon atoms is directed toward the vertices of the cube in which in the case of diamond carbon atoms are also located. These links are indicated in Figure 15 by thick dotted lines to mark the location of these connections is adamantine molecule. Solid thin lines delineate the regular tetrahedron inscribed in a cube. Its edges are the diagonals of the faces of the cube. Thick bar dotted lines delineate the regular octahedron, passing through the atoms located in the points $\alpha_2, \alpha_6, \alpha_3, \alpha_4, \alpha_8, \alpha_9$. The thin dash-dotted lines delineate the regular tetrahedron whose vertices coincide with the carbon atoms in a cube $\alpha_1, \alpha_5, \alpha_7, \alpha_{10}$.

By construction, the formed segments connecting the vertices of adamantane split into 10 families of parallel segments, each family including three parallel segments: (1) $\alpha_1\alpha_2, \alpha_7\alpha_6, \alpha_9\alpha_{10}$; (2) $\alpha_1\alpha_3, \alpha_8\alpha_{10}, \alpha_5\alpha_6$; (3) $\alpha_3\alpha_2, \alpha_5\alpha_7, \alpha_8\alpha_9$; (4) $\alpha_1\alpha_4, \alpha_7\alpha_8, \alpha_5\alpha_9$; (5) $\alpha_4\alpha_2, \alpha_6\alpha_8, \alpha_5\alpha_{10}$; (6) $\alpha_9\alpha_2, \alpha_3\alpha_8, \alpha_1\alpha_{10}$; (7) $\alpha_3\alpha_4, \alpha_7\alpha_{10}, \alpha_6\alpha_9$;

(8) $\alpha_2\alpha_5, \alpha_3\alpha_7, \alpha_4\alpha_{10}$; (9) $\alpha_1\alpha_5, \alpha_3\alpha_6, \alpha_4\alpha_9$; (10) $\alpha_2\alpha_6, \alpha_1\alpha_7, \alpha_4\alpha_8$. Consequently, the total number of segments (each of them is an edge of a polyhedron) is equal to 30. The length of segments is determined from the length of cube edges. Let's assume that the length of cube edge is equal to 1 (one should enter a scale factor to receive the specific dimension of a bond length). Then regular tetrahedrons with the bases on the faces of octahedron and the vertices coinciding with cube vertices (for example tetrahedron $\beta_1\alpha_2\alpha_3\alpha_4$) have the length $a = \frac{1}{\sqrt{2}}$ and the radius of circles described around the tetrahedrons is $b = \frac{\sqrt{3}}{4}$ (the points $\alpha_2, \alpha_3, \alpha_4, \alpha_6, \alpha_8, \alpha_9$ are in the centers of cube faces). Therefore, segments 2, 5, 6, 7, 9, 10 have the length *a*, while the segments 1, 3, 4, 8 have the length *b*. Thus, the two-dimensional geometric elements involved in adamantane have as sides the segments with lengths *a* and *b*. One can define (Figure 15) that a set of two-dimensional faces belonging to adamantane form regular triangles with sides *a*, an isosceles triangle with the base *a* and two sides *b*, squares with sides *a* and rectangles with sides *a* and *b*. Among the regular triangles, there are 4 triangles located at the outer edge of adamantane ($\alpha_2\alpha_3\alpha_6, \alpha_2\alpha_4\alpha_9, \alpha_3\alpha_4\alpha_8, \alpha_6\alpha_8\alpha_9$) and 8 triangles located in the inner part of adamantane

$$(\alpha_1\alpha_5\alpha_{10}, \alpha_1\alpha_7\alpha_{10}, \alpha_1\alpha_7\alpha_5, \alpha_2\alpha_6\alpha_9, \alpha_2\alpha_3\alpha_4, \alpha_3\alpha_6\alpha_8, \alpha_4\alpha_8\alpha_9, \alpha_5\alpha_7\alpha_{10}).$$

Among the irregular triangles, there are 12 triangles located at the outer edge of adamantane

$$(\alpha_2\alpha_1\alpha_3, \alpha_1\alpha_2\alpha_4, \alpha_1\alpha_3\alpha_4, \alpha_2\alpha_5\alpha_6, \alpha_3\alpha_7\alpha_8, \alpha_3\alpha_6\alpha_7, \alpha_4\alpha_9\alpha_{10}, \alpha_4\alpha_8\alpha_{10}, \alpha_5\alpha_6\alpha_9, \alpha_6\alpha_7\alpha_8, \alpha_8\alpha_9\alpha_{10})$$

and 6 triangles located in the inner part of adamantane

$$(\alpha_1\alpha_4\alpha_{10}, \alpha_2\alpha_1\alpha_5, \alpha_5\alpha_6\alpha_7, \alpha_1\alpha_3\alpha_7, \alpha_5\alpha_9\alpha_{10}, \alpha_7\alpha_8\alpha_{10}).$$

Thus, there are in total 30 triangles in adamantane. In the inner part of adamantane there are three squares with side *a* as three sections of the octahedron ($\alpha_2\alpha_4\alpha_6\alpha_8, \alpha_2\alpha_3\alpha_8\alpha_9, \alpha_3\alpha_4\alpha_6\alpha_9$) and 12 parallelograms (Figure 15) with sides *a* and *b*

($\alpha_1\alpha_3\alpha_8\alpha_{10}, \alpha_1\alpha_2\alpha_9\alpha_{10}, \alpha_1\alpha_4\alpha_5\alpha_9, \alpha_1\alpha_4\alpha_7\alpha_8, \alpha_1\alpha_3\alpha_5\alpha_6, \alpha_1\alpha_2\alpha_6\alpha_7, \alpha_2\alpha_4\alpha_5\alpha_{10},$

$\alpha_2\alpha_3\alpha_5\alpha_7, \alpha_3\alpha_4\alpha_7\alpha_{10}, \alpha_5\alpha_6\alpha_8\alpha_{10}, \alpha_5\alpha_7\alpha_8\alpha_9, \alpha_6\alpha_7\alpha_9\alpha_{10},$

$\alpha_7\alpha_3\alpha_2\alpha_5, \alpha_7\alpha_3\alpha_4\alpha_{10}, \alpha_5\alpha_7\alpha_8\alpha_9, \alpha_6\alpha_7\alpha_9\alpha_{10}\alpha_9\alpha_{10}$

$\alpha_7\alpha_3\alpha_2\alpha_5, \alpha_7\alpha_3\alpha_4\alpha_{10}, \alpha_5\alpha_7\alpha_8\alpha_9, \alpha_6\alpha_7\alpha_9\alpha_{10}\alpha_9\alpha_{10}$).

These parallelograms are rectangles, as one can prove that the planes of irregular triangles, resting upon the sides of specified squares, are perpendicular to the planes of these squares. Indeed, let's cut up the adamantane by a plane passing, for example (see Figure 15), through the top α_2 and the edges $\alpha_1\alpha_2, \alpha_2\alpha_5$ (due to the symmetry of the octahedron and tetrahedron built on its edges, these edges lie in the same plane). This plane cuts up irregular triangles $\alpha_1\alpha_3\alpha_4, \alpha_5\alpha_6\alpha_9$ and regular triangles $\alpha_2\alpha_3\alpha_4, \alpha_6\alpha_8\alpha_9$ at their heights, passing through the middle of the edges $\alpha_3\alpha_4, \alpha_6\alpha_9$ (respectively the points A_1, A_2 in Figure 15) and vertex α_8 . Intersection plane is presented in Figure 16.

Let's prove that the segments $\alpha_1 A_1$, $\alpha_3 A_2$ are perpendicular to the line A_1A_2. This will prove that the planes of irregular triangles are perpendicular to the plane of the square $\alpha_6\alpha_3\alpha_9\alpha_4$. Let us consider the triangle $\alpha_2\alpha_3A_2$; in it

$$A_1A_2 = a, \alpha_2A_2 = b = a\frac{\sqrt{6}}{4}, A_2\alpha_3 = \frac{a}{2\sqrt{2}}.$$

Therefore

$$\cos\angle\alpha_2A_2\alpha_3 = \sqrt{\frac{2}{3}}, \sin\angle\alpha_2A_2\alpha_3 = \frac{1}{\sqrt{3}}.$$

From the triangle $A_1A_2\alpha_2$, one can see that

Figure 16. Section of adamantane

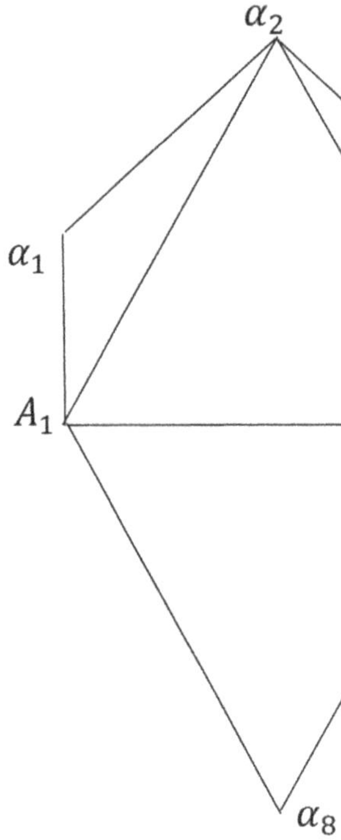

$$\cos \angle \alpha_2 A_2 A_1 = \frac{1}{\sqrt{3}}, \sin \angle \alpha_2 A_2 A_1 = \sqrt{\frac{2}{3}}.$$

Consequently:

$$\cos(\angle \alpha_2 A_2 \alpha_3 + \angle \alpha_2 A_2 A_1) = 0.$$

Then, $\alpha_3 A_2 \perp A_1 A_2$, QED.

This also implies that $\alpha_1\alpha_3 \perp \alpha_3\alpha_6$ and $\alpha_5\alpha_6 \perp \alpha_3\alpha_6$, in other words the parallelogram $\alpha_3\alpha_6\alpha_1\alpha_5$ is a rectangle. One can also prove that the remaining parallelograms are also rectangles. Thus, the number of squares and rectangles is 15 and the total number of geometric elements of dimension 2 consisting of adamantane is 45.

These 2*D* geometric elements form in adamantane 25 of 3*D* polyhedron (Figure 15):

- 5 tetrahedrons

$(\alpha_4\alpha_3\alpha_1\alpha_2, \alpha_{10}\alpha_7\alpha_1\alpha_5, \alpha_4\alpha_{10}\alpha_9\alpha_8, \alpha_3\alpha_6\alpha_7\alpha_8, \alpha_2\alpha_6\alpha_9\alpha_5)$,

- 6 prisms

$(\alpha_3\alpha_2\alpha_1\alpha_8\alpha_9\alpha_{10}, \alpha_6\alpha_2\alpha_1\alpha_7\alpha_9\alpha_{10}, \alpha_3\alpha_2\alpha_5\alpha_8\alpha_9\alpha_7, \alpha_3\alpha_5\alpha_1\alpha_8\alpha_9\alpha_{10},$
$\alpha_5\alpha_2\alpha_6\alpha_8\alpha_4\alpha_{10}, \alpha_5\alpha_4\alpha_1\alpha_8\alpha_9\alpha_7)$,

- 14 pyramids

$(\alpha_4\alpha_2\alpha_1\alpha_9\alpha_{11}, \alpha_5\alpha_2\alpha_1\alpha_9\alpha_{10}, \alpha_4\alpha_3\alpha_1\alpha_8\alpha_{10}, \alpha_7\alpha_3\alpha_1\alpha_8\alpha_{10}, \alpha_4\alpha_2\alpha_3\alpha_9\alpha_8,$
$\alpha_3\alpha_2\alpha_6\alpha_9\alpha_8, \alpha_3\alpha_2\alpha_1\alpha_5\alpha_6, \alpha_5\alpha_3\alpha_1\alpha_6\alpha_7, \alpha_4\alpha_3\alpha_6\alpha_9\alpha_8, \alpha_4\alpha_2\alpha_3\alpha_9\alpha_6,$
$\alpha_4\alpha_3\alpha_1\alpha_7\alpha_{10}, \alpha_4\alpha_3\alpha_7\alpha_8\alpha_{10}, \alpha_4\alpha_5\alpha_1\alpha_9\alpha_{10}, \alpha_4\alpha_2\alpha_1\alpha_9\alpha_5$.

When calculating 3*D* figures octahedrons as the figures consisting from two pyramids were not considered, because square sections of octahedron are involved in the formation of other 3*D* figures. Let's now calculate Euler - Poincare's formula (2) in Chapter 1 for the polytope *P* of dimension *n*.

As previously was defined in this case we have

$$f_0(P) = 10, f_1(P) = 30, f_2(P) = 45, f_3(P) = 25.$$

There are no elements of dimension greater than 3 inside adamantane. Substituting the values obtained for the number of faces of different dimension in Euler - Poincare's formula, for $n = 4$, we obtain $10 – 30 + 45 - 25 = 0$, i. e. Euler-Poincare's formula for adamantane is true at $n = 4$. This proves statement of theorem 6.

Adamantane is irregular convex polytope of dimension 4. From each vertex of this polytope outgoing 6 edges as in the 16-cell convex regular 4*D*-polytope (Grunbaum, 1967; Zhizhin, 2014a). All two-dimensional faces of adamantane are simultaneously the faces of two or more three-dimensional figures, which indicates the closeness of adamantane as a polytope. The existence of the outer three-dimensional adamantane boundary consisting of two-dimensional faces doesn't contradict to adamantane four-dimensionality if we take into account the inner structure. Just as the above-mentioned four-dimensional 4-crosspolytope can be considered as a figure consisting of two three-dimensional hexagonal pyramids applied to each other by their bases (Zhizhin, 2014a). Only drawing inside this figure six edges which form two regular triangles makes this figure a four-dimensional polytope consisting of 16 tetrahedrons. The outer boundary of adamantane consisting of two-dimensional faces of the polytope is the projection of the polytope on the three-dimensional space, just as the outer boundary of any closed polytope on a two-dimensional plane is a closed circuit composed of one-dimensional segments.

Silicon, germanium, tin and lead also form a molecule of adamantane or a molecule that is topologically close to the adamantane molecule (Gillespie, 1972). The dimension of these molecules is 4 or more than four. Phosphorus, antimony, arsenic and bismuth form already considered polytopes for other elements: a tetrahedron with a center, an octahedron with a center, an adamantane molecule. All these molecules have a dimensionality of 4 or higher. Chlorine, bromine and iodine show the greatest possible numbers of oxidation states, interacting with previously considered elements. The dimensions of these compounds are often higher than three. Halogen compounds lead to molecules of higher dimensionality. For example, the iodine heptafluoride molecule IF_7 has the form of a 6-simplex with a center whose dimension is 7. Noble gases in conjunction with other elements form molecules with tetrahedral coordination of higher dimension.

In the conclusion of the chapter, we give the table of the most common of binary compounds with *s* - and *p*- elements, indicating the type of structures of the molecules they form. The dimension of these molecules was determined earlier in the text of the Chapter 1 and Chapter 2.

Table 1. Binary compounds of the s- and p-elements

N	Type of the Structure	The Compounds Transition Elements With This Type of the Structure
1	rock salt	LiF, NaF, KF, LiCl, NaCl, KCl, RbCl, LiBr, NaBr, KBr, RbBr, LiI, NaI, KI, MgO, CaO, SrO, BaO, MgS, CaS, SrS, BaS, PbS
2	Adamantane	Pb_4O_6, As_4O_6, Sb_4O_6, P_4O_{10}, P_4O_4, $SiCl_4$, BeF_2, HgS
3	titanium chloride	RbF, CsF, CsCl, CsBr, CsI
4	Rutile	MgF_2, SnO_2, PbO_2
5	Wurtzite	BeO, AlN
6	Fluorite	CaF_2, SrF_2, BaF_2, RbF_2, AlF_8, $MgCl_2$, $CaCl_2$, $SrCl_2$, $BaCl_2$, $PbCl_2$, $AlCl_3$, $SiCl_4$, Li_2O, Na_2O, K_2O, Rb_2O, Cs_2O, Li_2S, Na_2S, K_2S, Rb_2S

REFERENCES

Baumschliter, C. W. (2007). Electronic and vibrational spectroscopy od diamondoids and the interstellar infrared bands between 3.35 and 3.55 μm. *The Astrophysical Journal*, *671*(1), 458–469. doi:10.1086/522683

Coxeter, H.S.M. (1963). *Regular Polytopes*. New York: John Wile & Sons, Inc.

Dahl, J. E., Liu, S. G., & Carlson, R. M. K. (2003). Isolation and structures of higher diamondoids, nanometer-sized diamond molecules. *Science*, *229*(5603), 96–99. doi:10.1126/science.1078239 PMID:12459548

Fisher, J., Baumgartner, J., & Marschner, C. (2005). Synthesis and structure of sila-adamantane. *Science*, *310*(5749), 825–830. doi:10.1126/science.1118981 PMID:16272116

Gillespie, R. J. (1972). *Molecular Geometry*. New York: Van Nostrand Reinhold Company.

Gillespie, R. J., & Hargittai, I. (1991). *The VSEPR Model of Molecular Geometry*. London: Allyn & Bacon.

Grunbaum, B. (1967). *Convex Polytopes*. London: Springer.

Hidaka, T. (1976). Bond-ioniciti anisotropy of the distorted wartzite-tipe crystals. *Physica. B+C*, *84*(30), 345–352. doi:10.1016/0378-4363(76)90043-7

Landa, S., & Machacek, V. (1933). Sur ladamantane, nouvel hydrocarbure extait du naphte. *Collection Czech Commun*, *5*, 1–5. doi:10.1135/cccc19330001

Sholl, C. A., & Walter, J. A. (1969). Covalent and ionic models for the electric field gradient in BeO. *Journal of Physical Chemistry*, *2*, 2301–2309.

Zhang, W., Oganov, A. R., Goncharov, A. F., Zhu, Q., Boulfelfel, S. E., Lyakhov, A. O., & Konopkova, Z. et al. (2013). Unexpected Stable Stereochemical of Sodium Chlorides. *Science*, *342*(6165), 1502–1505. doi:10.1126/science.1244989 PMID:24357316

Zhizhin, G. V. (2014a). On the higher dimension in nature. *Biosphere*, *6*(4), 313–318.

Zhizhin, G. V. (2014b). *World – 4D*. St. Petersburg: Polytechnic Service.

Zhizhin, G. V. (2016a). From the multidimensional phase space to a multidimensional space in nature. *Biosphere*, *8*(3), 258–267.

Zhizhin, G. V., & Diudea, M. V. (2016). Space of Nanoworld. In M. V. Putz & M. C. Mirica (Eds.), *Sustainable Nanosystems, Development, Properties, and Applications* (pp. 214–236). New York: IGI Global.

Zhizhin, G. V., Khalaj, Z., & Diudea, M. V. (2016). Geometrical and topology dimensions of the diamond. In A. R. Ashrafi & M. V. Diudea (Eds.), *Distance, symmetry and topology in carbon nanomaterials* (pp. 167–188). New York: Springer. doi:10.1007/978-3-319-31584-3_12

Zhou, X. F. (2012). First Principles Determination of the Structure of Magnesium Borohydride. American Physical Society, 109, 245503-1 – 245503-5.

Zhu, Q., Oganov, A. R., & Lyakhov, A. O. (2013). Novel stable compounds in the Mg – O system under high pressure. *Phys. Chem.*, *15*, 7696–7700. PMID:23595296

KEY TERMS AND DEFINITIONS

N-Cross-Polytope: The convex polytope of dimension n in which opposite related of centrum edges not have connection of edge.

N–Simplex: The convex polytope of dimension n in which each vertex is joined by edges with all remain vertices of polytope.

s- and p-Elements: The chemical elements in which is filling with electrons s- and p-orbitals of atoms.

Tetrahedral Coordination of Electron Pairs: The location of the electronic pairs of the outer and the pre-outer electron layer at the vertices of the tetrahedron.

The Divided Electron Pair: The binding electron pair, which simultaneously belongs to two atoms in the molecule.

Undivided Electron Pair: A non-bonding electron pair belonging to one atom in a molecule.

Chapter 3
The Structure, Topological, and Functional Dimension of Carbohydrates, Proteins, Nucleic Acids, ATP

ABSTRACT

New structures of biomolecules have been constructed: carbohydrates, proteins, nucleic acids. It is shown that glucose molecules and ribose molecules have dimensions of 15 and 12, respectively. The enantiomorphic forms of biomolecules in space of higher dimension make it possible to explain the experimentally observed facts of branching of chains of biomolecules in one of the enantiomorphic forms and the absence of chain branching in another enantiomorphic form. The enantiomorphic forms of the tartaric acid molecule in a space of higher dimension reveal the cause of the reversal in different directions of the polarization plane of light in two opposite forms.

FROM MULTIDIMENSIONAL PHASE SPACES OF DYNAMICAL SYSTEM TO MULTIDIMENSIONALITY OF BIOMOLECULES

The author has for many years studied the phase spaces of various dynamical systems: stationary flows of compressible media (Vooleys, Gusika & Zhizhin, 1971, 1972; Vooleys, Harachka & Zhizhin, 1977; Zhizhin, 1972, 1977, 2004a),

DOI: 10.4018/978-1-5225-4108-0.ch003

flow of incompressible media (Zhizhin & Ufimtsev, 1977, 1978; Zhizhin, 1980, 1987; Zhizhin & Onattsky, 1981), flow in reactors with chemical reactions (Gusika & Zhizhin, 1980; Zhizhin, 1980, 1984, 1988; Zhizhin & Segal, 1985, 1986). In these studies, as a rule, it was necessary to consider the multidimensional phase space of systems of differential equations, to investigate their special and singular points and the structure of the phase space as a whole. This led to an analysis of the spatial inhomogeneity that arise when the components of the medium interact under the influence of external influences. In particular this led to the formation of standing and traveling waves of chemical reactions: polymerization reactions (Zhizhin, 1982, 1984, 1985, 1992, 1997 a, b, c, 2000; Zhizhin et al., 1986 a, b; Zhizhin & Segal, 1986, 1988 a, b; Zhizhin & Obukhova, 1997), gas chemical reactions and detonation waves (Zhizhin & Larina, 1994; Zhizhin, 2005, 2008, 2009), waves of chemical reactions in condensed media (Zhizhin & Poritskaya, 1994; Zhizhin & Obukhova, 1995; Zhizhin, 2004 b). The interest in the formation of structures leads to the study of systems of differential equations and their phase spaces describing various processes in nature: the formation of "veins" of minerals (Zhizhin, 2004 c, d), waves of biological populations (Zhizhin & Bolshakova, 2000; Zhizhin, 2004 e, 2005 a, b, c), harvest programming (Zhizhin, 2011), formation of a large-scale structure of the Universe (Zhizhin, 2008), motion of prominences on the Sun (Zhizhin, 2010).

The study of the inhomogeneity in space naturally leads to an analysis of this inhomogeneity themselves and the determination of their structure. So there was accomplished a transition to the study of molecular structures. It was found that molecules can have a dimension greater than three (Zhizhin, 2014 a, b; Zhizhin, 2015, 2016a; Zhizhin & Diudea, 2016; Zhizhin, Khalaj & Diudea, 2016). In Chapters 1 and Chapter 2 it is clearly shown that the dimension of many molecules with the participation of elements of the periodic system of Mendeleev is higher than three. It is necessary to proceed to the study of the structure of biomolecules with the definition of their dimensionality, since their significance for life can not be overestimated.

Investigation of the structure of biomolecules is subject of many studies (bibliography on this subject can be found, for example, Metzler, 1980; Lehninger, 1982). These studies as a rule are based on the concept of the dimension of the objects considered no more than 3. In this context, and given the increased interest in the problems of nanomedicine, in this paper it is investigated the question of the dimension of biomolecules. For biomolecules are as small molecules called monomers (monosaccharides, amino acids, nucleotides) and biopolymers (proteins, polysaccharides, polynucleotides), consisting of a large (and often very high) of the corresponding monomers. Biomolecules are composed of carbon, nitrogen, oxygen, phosphorus, sulfur and small amounts of metal ions. The presence of these elements in biomolecules (as will be seen below) is essential to the determination of their dimension. The structure biopolymer quite difficult. In addition to the formation of the polymer chain, which is regarded as the primary structure, there are another three-level structure. The secondary structure describes a method for laying a polymer chain. The tertiary structure takes into account the interaction between circuit elements distant from each other. The quaternary structure describes the packaging of macromolecules in macroscopic formations. In this paper it is examine of the dimension of monomer units of polypeptide chains with regard to their atomic structure and functional groups within the monomers. Thus, attention is drawn primarily to the primary structure of its molecules, given the considerable importance of this structure when considering molecular structure of the following levels.

HIGHER DIMENSION OF POLYATOMIC MOLECULES AS A RESULT OF THE INTERACTION OF THE ELECTRON ORBITALS OF ATOMS IN A MOLECULE

The most common in biomolecules is a carbon atom, the main role of which to be binding in the center of biomolecules. Consider, for example, methane molecule CH_4. The carbon atom in this molecule binds around four hydrogen atoms. Geometrically, this molecule is a tetrahedron, whose vertices are located of the hydrogen atoms, and in the center is carbon atom. In Chapter 1 it is shown that the dimension of the molecules having the form of a tetrahedron with the center equals 4.

The carbon atom in the center of the methane molecule has the valence electrons $2s2p^3$. Valence electron orbitals of carbon atoms and hydrogen atoms $1s$ overlap and form four hybrid orbitals sp^3, directed from the carbon atom to the hydrogen atoms (Gray, 1965). If the distance from hydrogen atoms to carbon atoms is taken as unity, for the origin of coordinates to take the carbon atom, the directions hybrid orbitals send on four coordinates x, y, z, t, then the coordinates of the hydrogen atoms equal to (0, 0, 0, 1), (0, 0, 1, 0), (1, 0, 0, 0), (0, 1, 0, 0), and the carbon atom coordinates equal to (0, 0, 0, 0). So we have the integer coordinates of vertex in the four-dimensional space (Figure 1).

Figure 1. The methane molecule (CH_4)

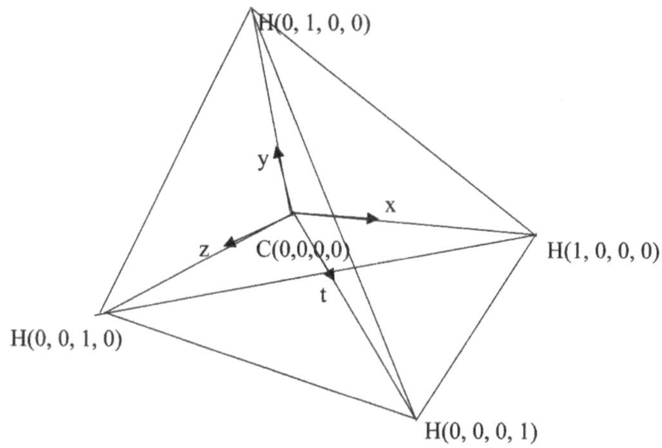

This is consistent with the evidence of four-dimension convex hull of the methane molecule on the Euler – Poincare equation (2) in Chapter 1. It is easy to see that the body in Figure 1, seen in the four-dimensional space, convex, because its edges belong to the body and enter into his boundary complex (Grunboum, 1967). Polytope in Figure 1 is a 4 - simplex, since each vertex of the polytope associated edges with all the other vertices of this polytope (Zhizhin, 2014 b).

If in the methane molecule a hydrogen atom replaced by a hydroxyl group - OH, then we get the simplest alcohol - methanol. If the hydroxyl group considered as the vertex of the polytope, then the dimension of this molecule will also be equal to 4. If each atom of the molecule of methanol is considered the vertex of the polytope, then connecting each vertex to all other vertices edges, it turns out that it is equal to the dimension of the polytope to 5 and we have 5 - simplex. However, here we must remember that the accession of the hydroxyl group does not change the hybridization of the carbon atom, as the binding site as the place of one hydrogen atom took one oxygen atom of the hydroxyl group. Therefore, as a separate vertex in methanol molecule should take hydroxyl group entirely. Then the dimension of the methanol molecules is equal to 4 (Zhizhin, 2017).

In the biomolecules can find a lot of examples of molecules or ions in the form of a tetrahedron with the center (NH_4 NH_4 , PO_4 PO_4 , etc.). All of them have dimension 4. If the binding site appears *d*- element it is formed around the coordination sphere ligands with more than 4 of the amount due to of *d*-orbitals of the element. One can show that in this case the dimension of the molecule is equal to the number of hybrid electron orbitals directed from the center to the ligands. The ligand may act as no individual atoms or ions, and some functional groups, which may be regarded as corresponding vertices of the polytope. This is consistent with the need to describe more convenient biomolecules, molecular structures consisting of different complexity. Therefore, the dimension of the group of atoms in biomolecules, we call a functional dimension. In addition, when such descriptions of specific dimensions we will not be considered distances between atoms in molecules. Therefore, a certain dimension of the molecules so called topological dimension.

In addition, of the tetrahedron with the center in biomolecules there are complex structures of higher dimension.

CARBOHYDRATE ("CARBON COMPOUND WITH WATER")

Carbohydrates are the main source of energy for the body. All carbohydrates are made up of units that are saccharides. The simplest saccharide is an aldose monosaccharide, which contains three carbon atoms (Figure 2).

The more complex monosaccharides include 4, 5, 6 and 7 carbon atoms. Polysaccharides consist of several monosaccharides. Saccharides are part of the nucleic acids that are carried out in the cells of protein synthesis and the transfer of hereditary traits. We will try to calculate the dimension of the simplest aldose monosaccharide saccharide, since the main elements of

Figure 2. Shema of the molecule aldose monosaccharide

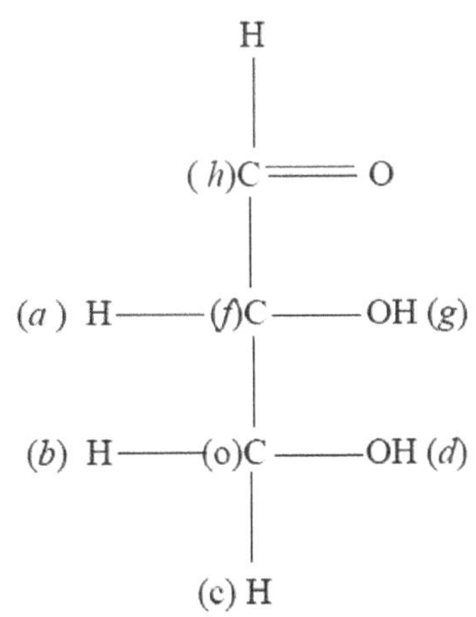

the aldose monosaccharide construction are repeated in the more complex saccharides. From

Figure 3. it follows that two carbon atoms, connected by bonds to each other, hydrogen atoms projection onto the plane, this construction is shown in Figure 3.

In this figure there is a tetrahedron *bcdf* with center *a* and a tetrahedron *oahg* with center *f*. Each a tetrahedron with a center is a polytope of dimension 4.

The vertex *f* of the first tetrahedron is the center of the second tetrahedron, and the vertex *o* of the second tetrahedron is the center of the first tetrahedron. Hydrogen (H) atoms are located at the vertices *a, b, c*, hydroxyl groups (OH) are located at the vertices *g, d*, carbon atoms (C) are located at the vertices *o, f, h*. The oxygen and hydrogen atoms following the carbon atom at the vertex *h* are not shown in Figure 3 for simplification. It is necessary to determine the dimension of the polytope *bcdfahgo*. The polytope in Figure 3 has 8 vertices (f_0 = 8), 22 edges (*ab, ag, af, ao, ah, gh, gf, go, gd, bf, bo, bd, bc, df, do, dc, hc, hf, ho, cf, co, fo*). Therefore, f_1 = 22. The polytope in Figure 3 has 29 planar faces, of which 26 are the triangles (*aho, afo, ahf, afg, ahg, aog, aob, afb, bfo, bco, bod, bfd, bfc, bcd, ghf, gho, gfo, god, gfd, dfo, dco, dfc, cof, chf, cho, hfo*) and 3 quadrangles (*abdg, hcgd, abhc*). Therefore, f_2 = 29. The polytope in Figure 3 has 20 three-dimensional figures, of which 13 are tetrahedrons (*bcfd, ahog, dcfo, bcdo, bfdo, cfdo, ahof, ahgf, hogf, aogf,*

Figure 3. Spatial structure of the molecule aldose monosaccharide

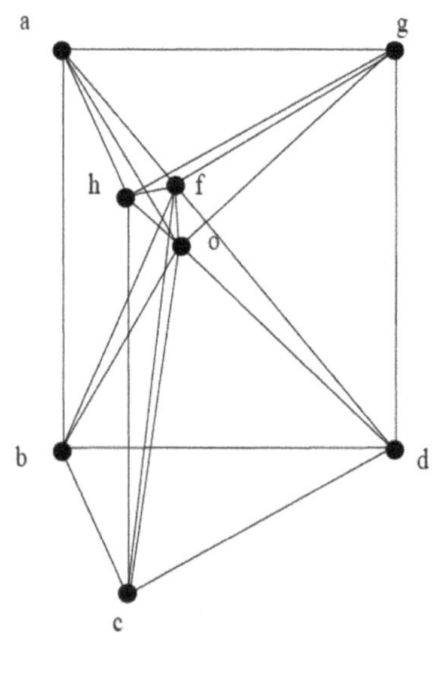

fgod, fohc, foab), 6 pyramids (*agbdo, ahbcf, ahbco, agbdf, Chgdf, chgdo*) and one (*ahgbcd)* prism. Therefore, $f_3 = 20$. It follows from the construction of the polytope in Figure 3 that it includes two tetrahedrons with the center *bcdfo* and *oahgf*. In addition to these two polytopes with dimension 4, five 4 - polytopes also appear in the polytope in Figure 3. Three of these polytopes have as their base three rectangular faces of the prism *ahgbcd*, whose vertices are connected with the vertices *f, o* located inside the prism. To prove their 4-dimensionality, consider one of these polytopes *abhcfo*, since the proofs for the other two polytopes are similar. This polytope has 6 vertices (f_0 = 6); 13 edges (*ab, ah, hc, bc, af, hf, bf, cf, ho, ao, bo, co, fo*), f_1 = 13; 13 two-dimensional faces (*ahf, aho, abo, abf, afo, bfo, boc, ahbc*), f_2 = 13; 6 three-dimensional faces (*hfoc, abof, bfoc, afho, ahcbf, ahcbo*), f_3 = 13. Substituting the obtained values of the numbers of faces of different dimensions into equation (2) in Chapter 1, we find that equation (2) is satisfied for n = 4

$$6 - 13 + 13 - 6 = 0,$$

Which is proved by the 4-dimensionality of the polytope *abhcfo*.

The two polytopes of dimension 4 there are formed by the *ahgbcd* prism with the vertex *f* or *o* inside its. Consider the prism *ahgbcd* with the vertex *f* (the proof for the prism with vertex *o* is similar). The polytope *ahgbcdf* has 7 vertices, $f_0 = 7$; 15 edges (*ah, hg, ag, bd, bc, cd, ab, hc, gd, af, fh, fg, bf, fc, fd*), f_1 = 15; 14 two-dimensional faces (*ahg, bdc, ahf, hfg, afg, bfc, fcd, bfd, fhc, afb, fgd, ahbc, hcgd, agbd*), f_2 = 14; 6 three-dimensional faces (*ahgbcd, ahgf, bcdf, abdgf, hgcdf, ahbcf*), $f_3 = 6$. Substituting the values of the numbers of faces of various dimensions obtained for the polytope *ahgbcdf* into equation (2) in Chapter 1, we find that it is satisfied for $n = 4$

$$7 - 15 + 14 - 6 = 0.$$

This proves that the polytope *ahgbcdf* has a dimension of 4.

Thus, for the polytope in Figure 3 are $f_0 = 8$, $f_1 = 22$, $f_2 = 29$, $f_3 = 20$, $f_4 = 7$. Substituting these values into equation (2) in Chapter 1, we find that it is satisfied for $n = 5$

$$8 - 22 + 29 - 20 + 7 = 2.$$

This proves that the polytope in Figure 3 has dimension 5. Consequently, the main part of the molecule of the aldose monosaccharide also has dimension 5. Since this basic part enters into all other saccharides in the plural, their dimension is more than 5. This gives higher dimension to the molecules DNA and all carbohydrates.

There are three main classes of carbohydrates: monosaccharides, oligosaccharides, polysaccharides. The basis of the monosaccharide is an unbranched chain of the carbon atoms, connected to each other by single bonds. One of the carbon atoms has double bond to an oxygen atom to form a carbonyl group. All other carbon atoms bonded hydroxyl groups and hydrogen ions. The carbonyl group may be at the end of the carbon chain (aldose) or elsewhere (ketoses). Monosaccharides depict a Fischer projection formula (Metzler, 1980; Lehninger, 1982). For example, the most common monosaccharides with five (pentose) and six (hexoses) carbon atoms in the form of these formulas are presented in Figure 4 and Figure 5 accordingly.

However, neither the Fischer formula or formula Haworth and their modifications (e.g., conformation as a "chair") may not reflect the spatial structure of the monosaccharides. For this target the constructs described in the form of convex polytopes with boundary elements which form boundary complex (Grunbaum, 1967). Only when such a representation will be to determine the dimension of these molecules. Consider a molecule of α - D - glucose. Closing unbranched chain of carbon atoms of monosaccharides through an oxygen atom, considering functional groups vertices, connect the each vertex by edges with each other vertices, get polytope, depicted in Figure 6.

Figure 4. The molecule of D – ribose

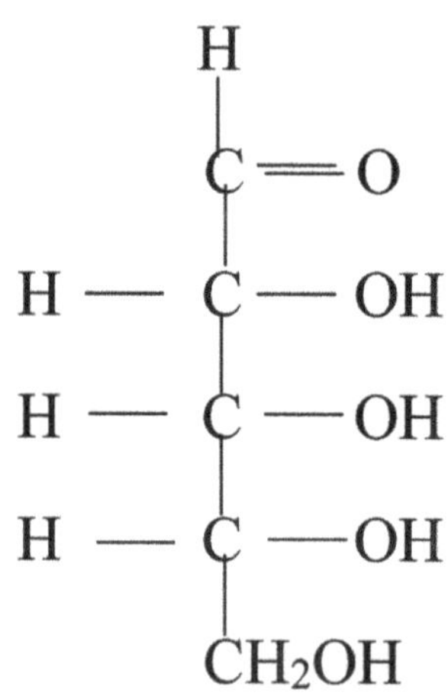

Figure 5. The molecule of D - glucose

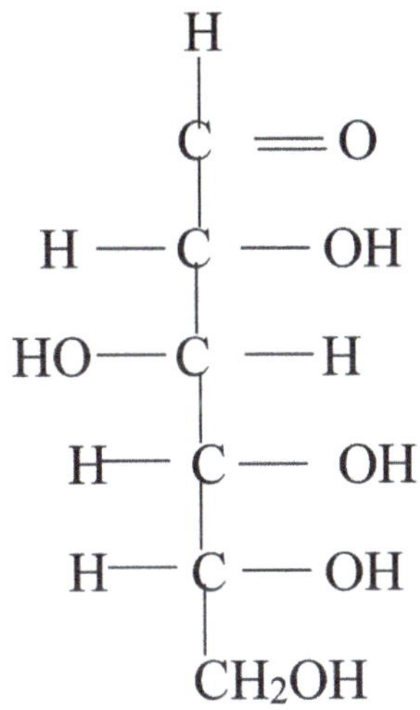

Figure 6. The molecule of α - D – glucose

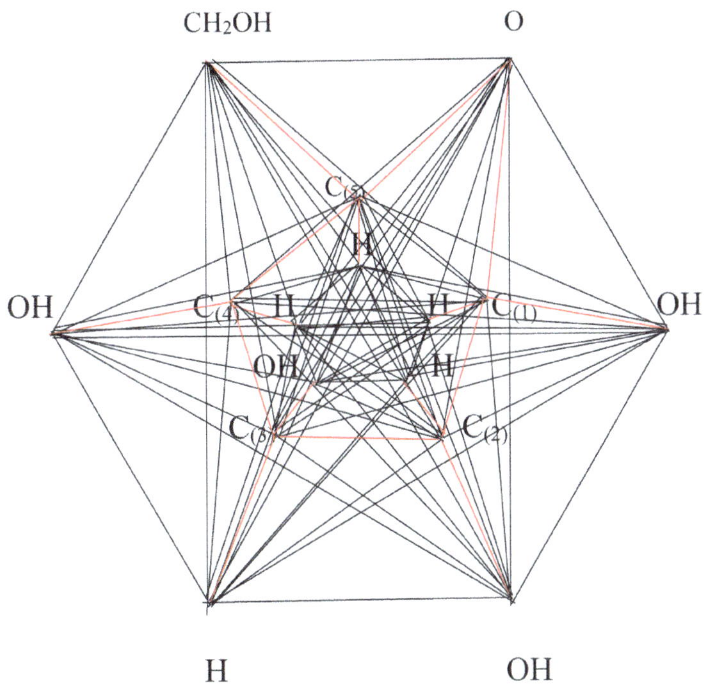

Edges marked in red in Figure 6 correspond to the chemical bonds in the molecule. The rest of the edges are only geometric sense, as the edges of the polytope.

Theorem 1

A molecule of α - D - glucose is a polytope type simplex with dimension 15 (Zhizhin, 2016 b).

Proof

The polytope in Figure 6 contains 16 vertices, i.e. $f_0 = 16 = n + 1$; 120 edges ($f_1 = C^2_{n+1} = 120$); 560 triangles ($f_2 = C^3_{n+1} = 560$); 1820 tetrahedrons ($f_3 = C^4_{n+1} = 1820$); 4368 $4D$ – simplexes ($f_4 = C^5_{n+1} = 4368$); 8008 $5D$ – simplexes ($f_5 = C^6_{n+1} = 8008$); 11440 $6D$ – simplexes ($f_6 = C^7_{n+1} = 11440$); 12870 $7D$ – simplexes ($f_7 = C^8_{n+1} = 12870$); 11440 $8D$ – simplexes ($f_8 = C^9_{n+1} = 11440$); 8008 $9D$ – simplexes ($f_9 = C^{10}_{n+1} = 8008$); 4368 $10D$ – simplexes ($f_{10} = C^{11}_{n+1} = 4368$); 1820 $11D$ – simplexes ($f_{11} = C^{12}_{n+1} = 1820$); 560 $12D$ – simplexes ($f_{12} = C^{13}_{n+1} = 560$); 120 $13D$ – simplexes ($f_{13} = C^{14}_{n+1} = 120$); 16 $14D$ – simplexes ($f_{14} = C^{15}_{n+1} = 16$).

Substituting the values $f_i, (0 \le i \le 15)$ in Euler's – Poincare equation (2) in Chapter 1, we see that it holds for $n = 15$

$$\sum_{i=0}^{14=n-1} f_i(-1)^i = 2.$$

This confirms that the polytope in Figure 6 has the dimension $n = 15$.

Theorem 1 it is proved.

High dimension of the molecule α - D – glucose is due to the fact that it contains a large number of differently oriented electronic atomic orbitals and, consequently, a large amount of energy. This is consistent with the established notions of large energy reserves in glucose, necessary for living organisms. Such an increase in energy and dimension occurs and other saccharides in the formation of closed loops.

In particular, the conformation of the β - D - glucose are interchanged only a hydroxyl group and a hydrogen atom bound to a carbon atom of the C (1) in Figure 6. Changes in the number of carbon atoms does not fundamentally change the picture of the molecule. The dimension of the polytope corresponds to the number of vertices of the polytope (not one less than the number of vertices).

Representations of the saccharide molecules in the form of polytope simplifies the understanding of the formation of polysaccharides. For example, if the molecule α - D - glucose, two molecules in accordance with Figure 6, are joined by the hydroxyl groups to form a water molecule and an oxygen atom, two molecules common α - D - glucose via α - glycoside linkages, as in simplified form shown in Figure 7.

Thus, the linear polymer of α - D - glucose has a one-dimensional translational symmetry with the translation element of higher dimension, just as quasicrystals (Shevchenko, Zhizhin & Mackay, 2013) have a multi-dimensional translational symmetry with the translation element of higher dimension. Chains with α - glycoside bond have the opportunity to branch. This is evident from Figure 5, as the functional group -CH_2OH in each molecule can be a chain branch point, to which is attached via an oxygen atom molecule of α - D - glucose. In the case of β - glycoside bond molecules β - D - glucose (Figure 8) such a possibility is difficult due to a denser arrangement of glucose molecules, and proximity to a functional group – CH_2OH of oxygen atom.

It seems can serve as an explanation of a chain with β - glycoside linkage chain branching is not observed.

Figure 7. The α - glycoside linkage of the molecules α – D – glucose

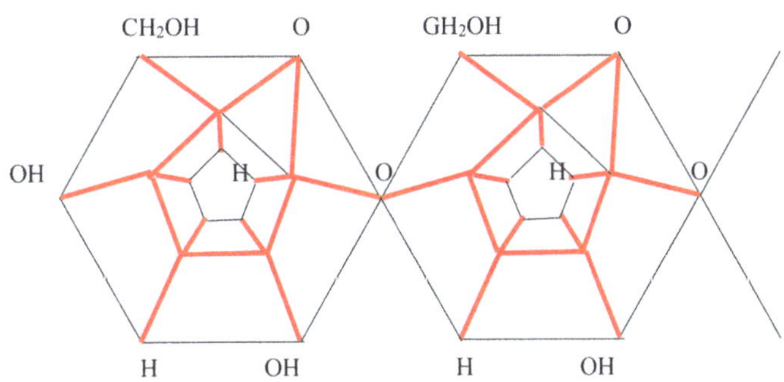

Figure 8. The β - glycoside linkage of molecules β - D - glucose

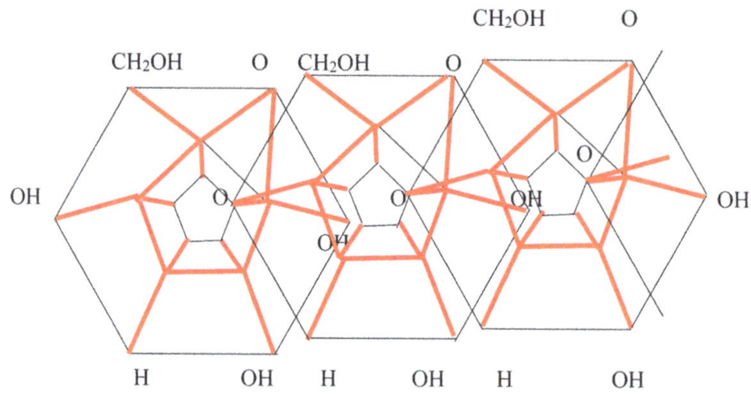

PROTEINS

Monomer units which are built of proteins are the 20 standard amino acids. These small molecules containing two different chemical functional groups capable of reacting with each other to form a covalent bond. This are amino group (-NH_2) and a carboxyl group (- COOH). Connection which determines the formation of protein polymer is called a peptide bond. In the formation of such a connection by joining together - COOH and -NH_2 with secretion a molecule of water. Amino acids forming two families of *D* and *L*, each of which can be represented in the form of a tetrahedron with the center in the carbon atom (Figure 9, Figure 10)

According to the ideas of this work, the amino acid is a molecular formation with topological and functional dimension of 4, regardless of the structure of the side of the functional groups represented by *R*.

Communication amino acids can be represented as a tetrahedron with the center of a peptide bond. Figure 8 shows the peptide bond the *D* - amino acids.

From geometric images of associated polytopes of dimension 4 in Figure 11 immediately implies that the peptide chain has the form of a spiral, swirling clockwise. Side functional groups *R* have different chemical nature. Sequence arrangement of functional groups and, hence, the sequence of amino acids in the chains always accurately defined genetically.

The peptide chain may form both parallel and antiparallel structure associated hydrogen bond. In addition, the peptide chain may form a compact protein globule. This class of proteins known as globular proteins that perform complex biological functions. For example, the protein is a globular myoglobin

Figure 9. The molecule of D - amino acid

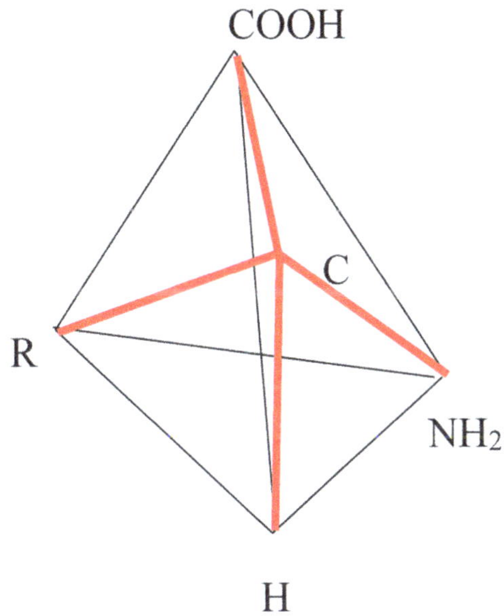

Figure 10. The molecule of L - amino acid

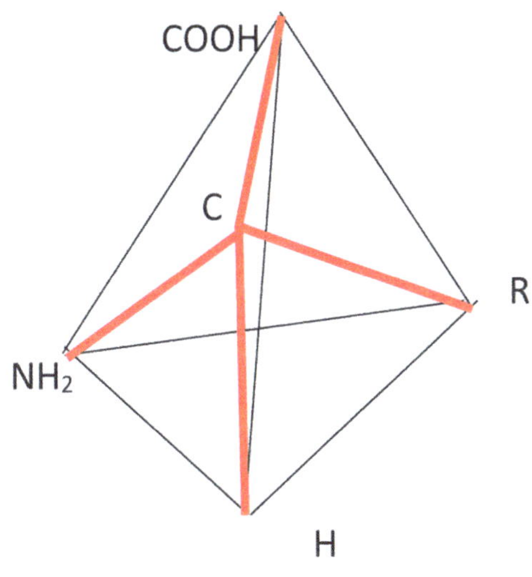

Figure 11. The peptide chain of amino acids

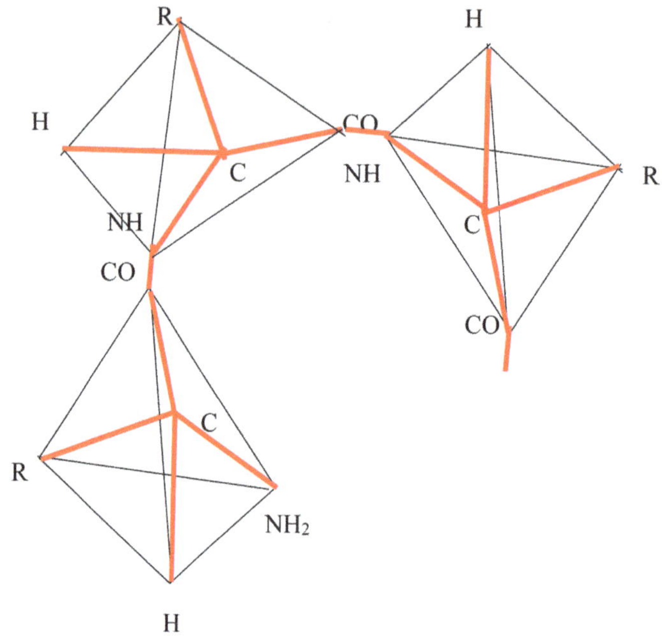

- oxygen-binding protein present in the muscles. In the center of myoglobin globule is hemo-group containing Fe - porphyrin (iron atom surrounded by five nitrogen atoms).

Theorem 2

The dimension of the Fe - porphyrin before joining the oxygen atom is equal to 5.

Proof.

Consider the first coordination sphere of the iron atom in the center of the porphyrin (Zhizhin, 2015), since only in the first coordination sphere of atoms are linked by a covalent bond, and in the following focal areas of intermolecular bonds between atoms. Before joining of the oxygen atom the first coordination sphere of Fe - porphyrin may be represented as a plane projection (Figure 12), at the vertices *a, c, d, f* of which the nitrogen atoms of the porphyrin are located, an iron atom is located at the vertex *g*, and the nitrogen atom of the nearest histidine residue is located at the vertex *b*. The

Figure 12. The first coordination sphere of Fe - porphyrin before binding oxygen

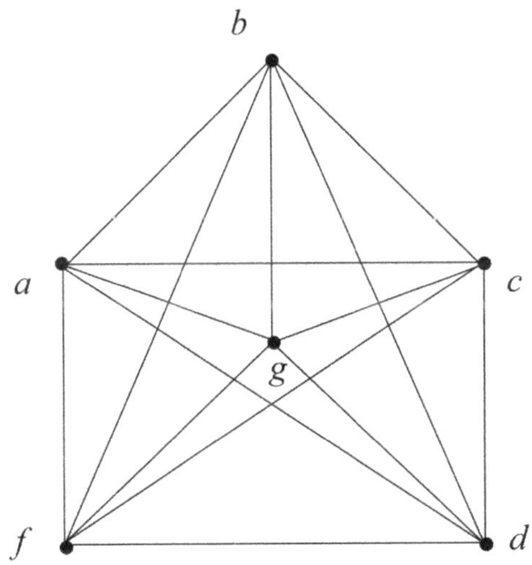

deflection of vertex *g* from the center of the rectangle *acdf* corresponds to a certain "dome" character of porphyrin (Steed & Atwwod, 2007 ; Lehn, 1998). The projection in Figure 12 represents some polytope (let`s denote *A* - polytope).

The *A* – polytope has six elements with dimension 0, $f_0(A) = 6$. There are vertices *a, c, d, f, g, b*. The number of elements with dimension 1 is $f_1(A) = C_6^2 = 15$. It are edges *ab, bc, bd, bf, bg, ac, cd, fd, af, fc, ad, ag, gc, fg*. The number of elements with dimension 2 is $f_2(A) = C_6^3 = 20$. It are triangles *abf, bfg, bgd, dbc, bga, bgc, agc, dfg, adc, acf, fcd, bgd, fbg, agd, fgc, fbc, abd, afg, gcd, afd*). The number of elements with dimension 3 is $f_3(A) = C_6^4 = 15$. It are tetrahedons *abgf, bsgd, abfc, abcd, bfcg, abdg, acfg, abdf, acdg, bfdg, abgc, fbcd, fgcd, afgd, afcd*). The number of elements with dimension 4 is $f_4(A) = C_6^4 = 6$. It are simplexes *abcdf, adcdg, abdfg, abcfg, bcdfg, acdfg*. Substituting the received numbers of elements of different dimensions in the equation (2) in Chapter 1 at a value of *n* = 5, we obtain

6 - 15 + 20 - 15 + 2 = 2,

i.e. the Euler- Poincare equation is satisfied for *A* - polytope with *n* = 5. This is a simplex of dimension 5. This proves theorem 2.

Theorem 3

The dimension of Fe - porphyrin after joining of oxygen atom is 6.

Proof

The first coordination sphere after joining oxygen atoms is complemented by one vertex *e* (Figure 13).

"Dome" character of Fe - porphyrin after joining of oxygen atom decreases, but it is not possible to affirm that it disappears completely (Steed & Atwood, 2007). Therefore, the deflection of vertex *g* from the center of rectangle in Figure 13 quality is maintained qualitatively. Taking into account the significant difference between the geometry and mass of the groups attached to the iron atom at the top and bottom, it is shown in Figure 13 that the vertices *e* and *b* do not lie on the same line. In the polytope in Figure 13 (*B* - polytope) the number of elements of zero dimension is increased compared with to the *A*-polytope by one vertex *e*, $f_0(B) = 7$. This leads to the increase in the dimension of the polytope by 1, as the number of edges issuing from each

Figure 13. The first coordination sphere of Fe - porphyrin after joining of oxygen atom

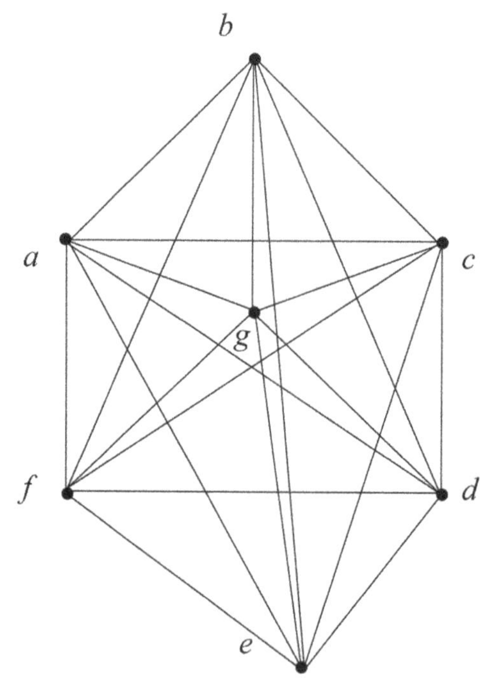

top also increases by 1. In the polytope *B* the number of elements of dimension 1 is $f_1(B) = C_7^2 = 21$ (edges). The number of elements with dimension 2 is $f_2(B) = C_7^3 = 35$ (triangles). The number of three-dimensional figures is $f_3(B) = C_7^4 = 35$ (tetrahedrons). The number of elements with dimension 4 is $f_4(B) = C_7^5 = 21$, (simplexes of dimension 4). The number of elements with dimension 5 is $f_5(B) = C_7^6 = 7$, (simplexes of dimension 5). Substituting the numbers of the elements of different dimensions in equation (2) in Chapter 1 with $n = 6$, we get

$$7 - 21 + 35 - 35 + 21 - 7 = 0,$$

i.e. the Euler - Poincare equation for *B* - polytope is satisfied when $n = 6$. Therefore, *B*- polytope is a simplex of dimension 6. This proves theorem 3.

The dimensions of molecules increase with an increase of its energy again. It is shown that myoglobin is associated coil circuit elements of higher dimension (4) and, moreover, in the center of the coil is a group of atoms even greater dimension.

NUCLEIC ACIDS, ATP

Nucleic acids (DNA and RNA) are polynucleotide. These monomer units (nucleotides) consists of pyrimidine and purine bases, *D*-ribose (or *D*-2-deoxyribose) and phosphoric acid. The bases are virtually flat molecules (Metzler, 1980; Lehninger, 1982), we will be denoted R_f. As follows from the analysis conducted in carbohydrates *D* - ribose molecule has a higher dimension. Phosphoric acid has the structure (Figure 14)

It follows from claim 1, which, apart from a double bond between phosphorus and oxygen atoms is edge polytope, phosphoric acid molecule a geometrically is tetrahedron with the center, thus it has dimension 4.

Let is present a molecule *D*-ribose as a polytope (Figure 15).

Theorem 4

The molecule of *D*-ribose is a convex polytope type simplex of dimension 12.

Figure 14. The structure of phosphoric acid

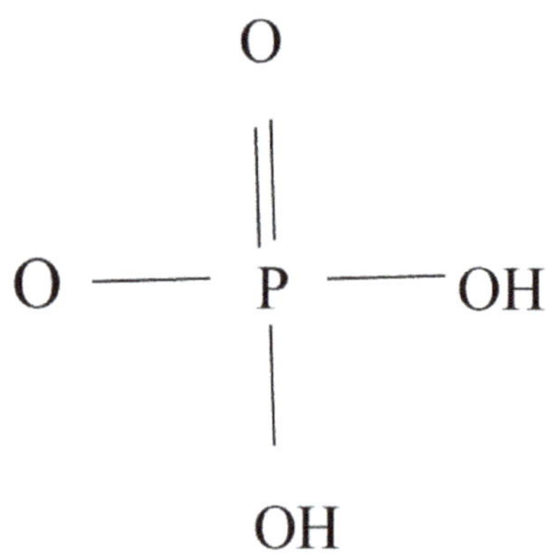

Figure 15. The molecule D-ribose

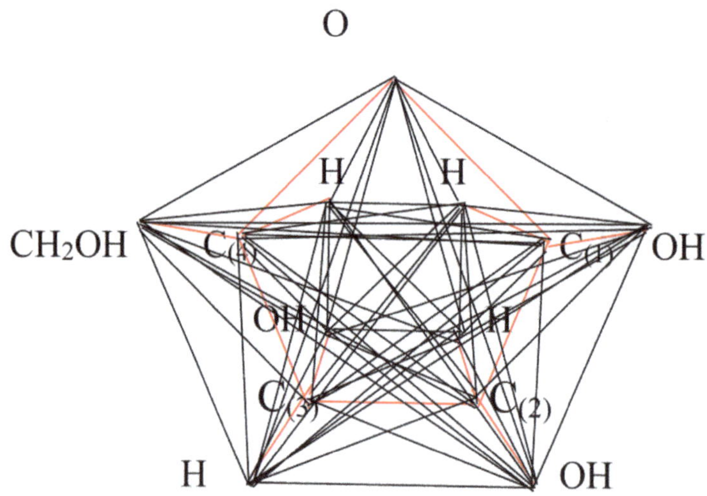

Proof

Polytope in Figure 15 has vertices 13, $f_0 = 13 - n + 1$; 78 edges ($f_1 = C^2_{n+1} = 78$); $f_1 = C^2_{n+1} = 78$); 560 triangles ($f_2 = C^3_{n+1} = 286$);7 $f_2 = C^3_{n+1} = 286$);7 15 tetrahedrons ($f_3 = C^4_{n+1} = 715$); 1287 4*D*-simplexes ($f_4 = C^5_{n+1} = 1287$); 1716 5*D*-simplexes ($f_5 = C^6_{n+1} = 1716$); 1716 6*D* – simplexes ($f_6 = C^7_{n+1} = 1716$); 1287 7*D*-simplexes ($f_7 = C^8_{n=1} = 1287$);715

$8D$-simplexes $(f_8 = C^9_{n+1} = 715); 286$ $9D$-simplexes $(f_9 = C^{10}_{n+1} = 286); 78$ $10D$-simplexes $(f_{10} = C^{11}_{n+1} = 78)$; 13 $11D$-simplexes $(f_{11} = C^{12}_{n+1} = 13)$.

Substituting the values $f_i, (0 \le i \le 11)$ in Euler's - Poincare equation (2) in Chapter 1, we see that it holds for n = 12

$$\sum_{i=0}^{11=n-1} f_i(-1)^i = 0.$$

Theorem 4 is proved.

The polynucleotides molecules D - ribose combined with phosphoric acid residues and bases R_f development of two water molecules (R_f have six kinds of alternate bases - pyrimidine, uracil, adenine, cytosine, guanine, purine). Thus, the polynucleotide chain is a sequence related to each other through oxygen atoms higher dimensional objects (simplexes)

4D - O - 12D - O - 4D - O - 12D - O - ....

Nucleotides ($12D$ - simplex) act in some cases as a coenzyme of biochemical reactions. For example, nucleotides associated with two extra residues in the form of phosphoric acid of polyphosphoric acid to form (Figure 16) adenosine triphosphate (ATP)

Compounds of this type are readily cleaved, one or two phosphoric acid residue, which is transferred to any other radicals, - the process of so-called phosphorylation. Bond in the chain polyphosphoric rich in energy, so simultaneously with the transfer of phosphorus is carried energy transfer from one connection to another. Thus, ATP is also a chain of elements of higher dimension (simplexes)

Figure 16. The scheme of the ATP

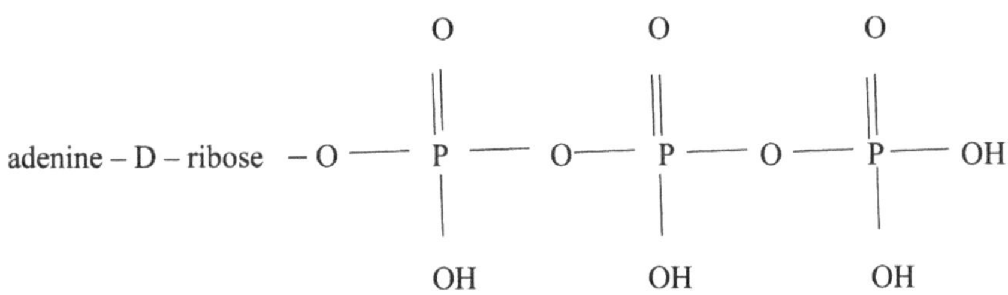

R_f - 12D - 4D - 4D - 4D.

In biomolecules are essential sense transition metals. They, being in the center of the coordination spheres, provide management role in the living organisms. This is achieved due to the presence of these metals in a large number of electrons and the quantum of vacant cells in the outer shell of atoms. Due to the transition metals carried covalent and donor-acceptor chemical bonds with atoms other elements in the living organisms. A significant part of the transition metals have a deviation from the rules of filling of electron orbitals in order of increasing energy falling on the orbitals of higher energy. Currently there is no classification and analysis of the anomalous transition metals having such deviations. Considering the importance of these metals to the functioning of living organisms, it is of interest for further work to examine these anomalies in order to establish operating in these patterns and identify opportunities for their practical use.

WINE ACID

Earlier in Chapter 3, structures of two enantiomorphism forms of glucose were considered. The construction of images of their molecules in a space of dimensionality 15 made it possible to explain why the chain of molecules of α-*D* glucose has branches from the chain, and the chain of molecules of the *β-D* glucose molecule does not have a branch from the chain. Earlier in Chapter 3, the structure of the aldose monosaccharide in the configuration of *D* is considered (Metzler, 1980). The aldose monosaccharide also has an enantiomorphism configuration of *L*. We will consider the difference in these configurations by the example of a closely related tartaric acid, which played a major role in the development of biology, starting with Pasteur's well-known works (Pasteur, 1960). However, instead of the known images of these molecules in the form of Fisher's projections (Figure 17, Figure 18), we will use images of space of higher dimension for their images.

Comparing Fisher images of aldose monosaccharide (Figure 2) and tartaric acid (Figure 17, Figure 18), we see that these compounds have the same main part of the design. It has the form of two tetrahedrons with a center, and the center of each of them is simultaneously the vertex of another tetrahedron. There is some difference in functional groups of compounds. Enantiomorphism forms of tartaric acid differ in the mirror image of hydrogen ions and hydroxyl

Figure 17. Shema of the molecules D-wine acid

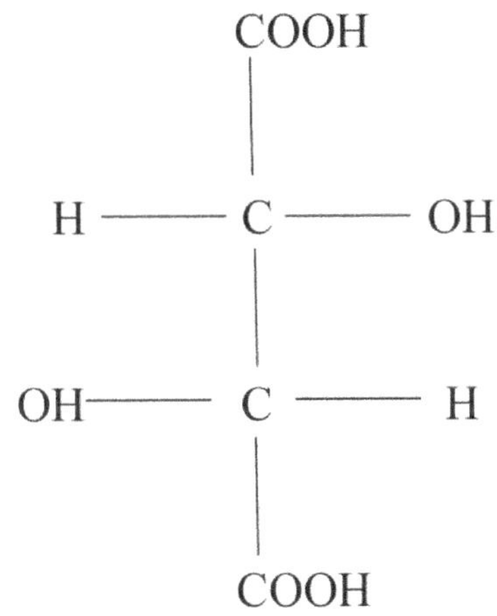

Figure 18. Shema of the molecules L-wine acid

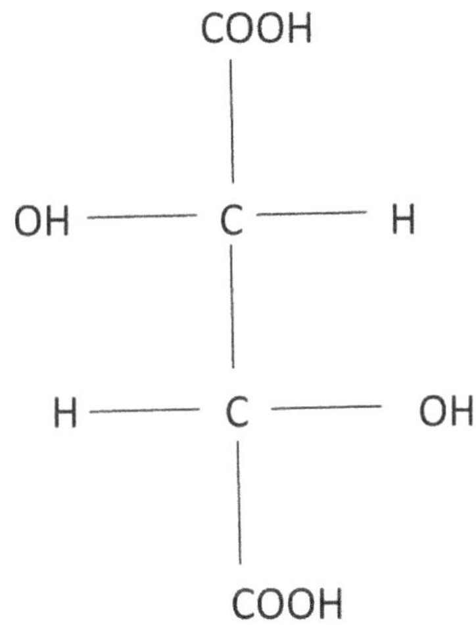

groups in the main part of the molecule's structure. The dimension of this construction how it is shown earlier equal 5. Thus, the dimension of the molecules tartaric acid in both forms is 5. Images of polytopes corresponding to a molecule of tartaric acid in the form *D* and form *L* are presented in Figures 19 and Figure 20.

The brown color in Figures 19, 20 denotes the edges corresponding to the chemical bonds between the atoms. The black color in these figures denotes the edges that have values only as the edges of the convex body. The outer contour

Figure 19. Spatial structure of the D-wine acid

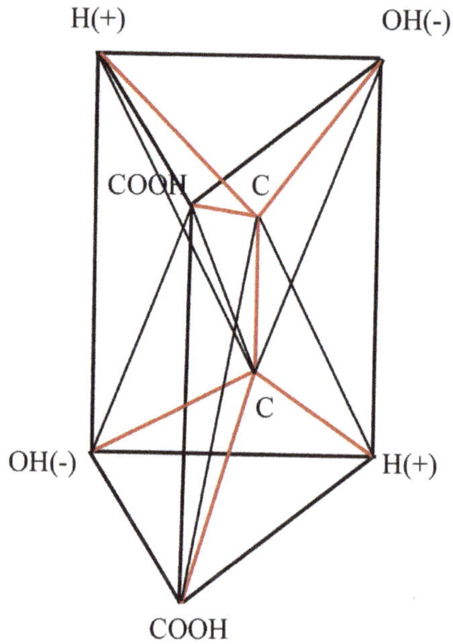

Figure 20. Spatial structure of the L-wine acid

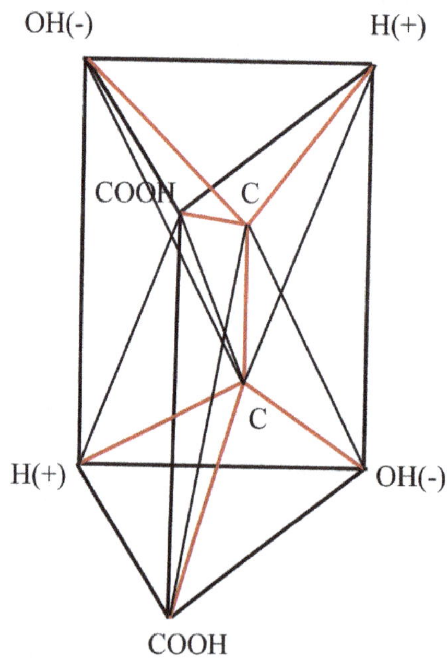

of both molecules in three-dimensional space is a triangular prism. There are two carbon atoms within these prisms. These two carbon atoms and lead to an increase in the dimension of the molecule to five. On the outer contour, two enantiomorphism forms have the opposite arrangement of hydrogen ions and a hydroxyl group. The images obtained make it possible to explain the main property of tartaric acid - rotation of the plane of polarization of the incident light in different directions: in the case of the *D* form to the right, in the case of the *L* form to the left. It are known devices for rotating the plane of polarization of light, having the appearance of two folded triangular prisms, the boundary between which serves to reflect light (Вуд, 1936). We can say that the molecule of tartaric acid is a natural device for rotating the plane of light polarization. Two carbon atoms play the role of the reflecting partition in the molecule. The rotation occurs in the forms *D* and *L* in different directions because of the opposite arrangement of the charges of the hydrogen ions (+) and the hydroxyl group (-) in these forms. Thus, the reason for the different rotation of the plane of polarization of light lies not in the different forms of the crystals of *D*-tartaric acid and *L*-tartaric acid, as Pasteur suggested, but in different forms of molecules, clearly visible in the image in the space 5*D*.

A number of serious works on the use of spaces of higher dimension in the analysis of the structure of viruses belongs to the authors Janner and Twarock (Janner, 2006, 2008, 2011, 2016; Keef & Twarock, 2009; Twarock & Dykeman, 2010). However, it should be noted, that in these works, especially in the works of Janner, the notion of polytopes of higher dimension is often used incorrectly. The quantities of elements of different dimensions are not determined and the feasibility of the Euler - Poincare equation is not checked. Therefore, the results of these studies require verification.

REFERENCES

Grunbaum, B. (1967). *Convex Polytopes*. London: Springer.

Gusika, P. L., & Zhizhin, G. V. (1980). Unsteady conditions of radical polymerization. *Applied Mathematics and Mechanics*, *4*, 702–709.

Janner, A. (2006). Towards a classification of icosahedral viruses in term of indexed polyhedral. *Acta Crystallographica. Section A, Foundations of Crystallography*, *62*(Pt5), 319–330. doi:10.1107/S0108767306022227 PMID:16926480

Janner, A. (2008). Comparative architecture of octahedral protein cages. II. Interplay between structural elements. *Acta Crystallographica. Section A, Foundations of Crystallography*, *64*(Pt4), 503–512. doi:10.1107/S0108767308012051 PMID:18560167

Janner, A. (2011). Form, symmetry and packing of biomacromolecules. Shells with boundaries at anti-nodes of resonant vibration in icosahedral RNA viruses. *Acta Crystallographica. Section A, Foundations of Crystallography*, *67*(Pt6), 521–532. doi:10.1107/S010876731103577X PMID:22011468

Janner, A. (2016). Symmetry-adapted digital modeling III, Coarse-grained icosahedral viruses. *Acta Crystallographica. Section A, Foundations of Crystallography*, *72*(Pt3), 324–337. doi:10.1107/S205327331600276X PMID:27126109

Keef, T., & Twarock, R. (2009). Affine extensions of the icosahedral group with applications to the three-dimensional organization of simple viruses. *Journal of Mathematical Biology*, *59*(3), 287–313. doi:10.1007/s00285-008-0228-5 PMID:18979101

Lehn, J. M. (1998). *Supramolecular chemistry. Concept and perspectives*. Novosibirsk: Science.

Lehninger, A. L. (1982). *Principles of Biochemistry* (Vol. 1-3). New York: Worth Publisher, Inc.

Metzler, D. E. (1980). *Biochemistry. The Chemical Reactions of Living Cells* (Vol. 1-3). New York: Academic Press.

Pasteur, L. (1960). *Selected works*. Moscow: Publishing House of the Academy of Sciences of the USSR.

Shevchenko, V. Ya., Zhizhin, G. V., & Mackay, A. (2013). On the Structure of Quasicrystals in a Higher-Dimensional Space. In M. V. Diudea & C. L. Nagy (Eds.), *Diamonds and related nanostructures* (pp. 311–320). New York: Springer. doi:10.1007/978-94-007-6371-5_17

Steed, J. V., & Atwood, J. L. (2000). *Supramolecular chemistry*. Chichester, UK: John Wiley and Sons.

Twarock, R., & Dykeman, E. C. (2010). Al-atom normal-mode analysis reveals an RNA-induced allostery in a bacteriophage coat protein. *Physical Review E: Statistical, Nonlinear, and Soft Matter Physics*, *81*(3), 1–10.

Vooleys, L. A., Gusika, P. L., & Zhizhin, G. V. (1971). A qualitative study of one-dimensional mounted MHD - flow conducting gas. *Magnetohydrodynamics*, *4*, 11–17.

Vooleys, L. A., Gusika, P. L., & Zhizhin, G. V. (1972). Two-phase flow in a channel of constant cross section (qualitative research). *Journal of Applied Mechanics and Technical Physics*, *5*, 143–156.

Vooleys, L. A., Harachka, I. G., & Zhizhin, G. V. (1977). *The equilibrium two-phase flow in tubes with variable flow. In Boiling crisis and the critical region* (pp. 112–117). Leningrad: Science.

Wood, R. (2016). *Researches in physical optics*. Wentworth Press.

Zhizhin, G. V. (1972). *A qualitative study of one-dimensional steady flows* (thesis abstract). Leningrad: Publishing House of Leningrad Polytechnic Institute.

Zhizhin, G.V. (1977). Two-phase flow with friction. *Engineering-Physics Journal, 1*, 186 - 8.

Zhizhin, G. V. (1980). Analysis of the flow reactor with ideal mixing in the presence of the three phases. *Kinetics and Catalysis*, *2*, 511–518.

Zhizhin, G. V. (1981). Isothermal Couette flow of non-Newtonian fluid under the influence of a pressure gradient. *Journal of Applied Mechanics and Technical Physics*, *3*, 26–30.

Zhizhin, G. V. (1982). The structure of the wave front of the polymerization. *Reports of the USSR Academy of Sciences*, *263*(6), 1399–1402.

Zhizhin, G. V. (1984). Research reactor operation semi-continuous. *Theoretical Foundations of Chemical Engineering*, *6*, 769–774.

Zhizhin, G. V. (1984). Calculation of wave structure of the frontal polymerization. *Kinetics and Catalysis*, *25*(2), 292–298.

Zhizhin, G. V. (1985). Determination of coordinates of the front-radical polymerization. *Kinetics and Catalysis*, *4*, 1002–1005.

Zhizhin, G. V. (1986). Stationary states of a spherical reaction front in a liquid with variable viscosity. *Kinetics and Catalysis*, *6*, 1310–1314.

Zhizhin, G. V. (1986). On peculiarities of propagation of the heat wave in the thermal polymerization of vinyl monomers. *Chemical Physics*, *10*, 1421–1425.

Zhizhin, G. V. (1987). The laminar boundary layer of non-Newtonian fluids (qualitative research). *Journal of Applied Mechanics and Technical Physics, 3*, 71–81.

Zhizhin, G. V. (1988). Autowave processes of distribution of chemical reactions in a dispersion medium. *Journal of Applied Mechanics and Technical Physics, 6*, 35–43.

Zhizhin, G. V. (1992). *Macrokinetics of frontal polymerization reactors*. St. Petersburg: Polytechnica.

Zhizhin, G. V. (1997). Stationary waves reversible radical polymerization. *Chemical Physics, 3*, 114–123.

Zhizhin, G. V. (1997). *Mathematical models of waves radical polymerization*. St. Petersburg: Publishing Northwestern Correspondence Polytechnic Institute.

Zhizhin, G. V. (1997). Stationary Waves of Reversible Radical Polymeryzation. *Chem. Phys. Reports, 16*, 515–526.

Zhizhin, G. V. (2000). Mathematical model of isothermal diffusion wave of radical polymerization. *Chemical Physics, 10*, 72–78.

Zhizhin G.V. (2004a). Synergetics of flows. *The Life and Safety, 5-6,* 520 - 550.

Zhizhin, G. V. (2004b). The model of an ideal wave hard-flame combustion variable surface chemical interactions. *Combust, 1*, 95–102.

Zhizhin, G. V. (2004c). Modeling Liesegang bands formed at filtration chemically active liquids. *Chemical Physics, 1*, 82–89.

Zhizhin, G. V. (2004d). Modeling Liesegang bands formed in the ion exchange reaction. *Chemical Physics, 10*, 54–61.

Zhizhin, G. V. (2004e). *Self-regulating wave of chemical reactions and biological populations*. St. Petersburg: Science.

Zhizhin, G. V. (2005). Still dissipative structures formed by soil bacteria. *Biophysics, 2*, 322–328.

Zhizhin, G. V. (2005). *Dissipative structures in chemical, geological and ecological systems*. St. Petersburg: Science.

Zhizhin, G. V. (2005). Immobile Dissipative Structures by Soil Bacteria. *Biophycics, 50*(2), 303–308.

Zhizhin, G. V. (2008). *Combustion waves with distributed zones of chemical reactions (nonasymptotic combustion theory)*. St. Petersburg: Publishing Werner Regen.

Zhizhin, G. V. (2008). To a question on possible mesh large-scale structure of the Universe. Practical cosmology. In *Proceedings of International conference "Problems of practical cosmology"* (pp. 134-6). St. Petersburg: Russian Geographical Society.

Zhizhin G.V. (2009). *Detonation waves in a gas with variable composition*. St. Petersburg: Polytechnic-Service.

Zhizhin, G. V. (2010). *Geometric foundations of dissipative structures*. St. Petersburg: Polytechnica.

Zhizhin, G. V. (2010). A qualitative study of motion of the gas in the prominences on the Sun. In Jet, separated and unsteady flow (pp. 74-76). St. Petersburg: Publishing Baltic State Technical University.

Zhizhin, G. V. (2011). The method of programming tasks in high yield. In Actual problems of biology and ecology (pp. 133-142). St. Petersburg: Publishing St. Petersburg Forest Technical Academy.

Zhizhin, G. V. (2014b). On the higher dimensions in nature. *Biosphere*, *6*(4), 313–318.

Zhizhin, G. V. (2015). The dimensions of supramolecular compounds. *Biosphere*, *7*(2), 149–154.

Zhizhin, G. V. (2016a). From multidimensional phase spaces to multidimensionality of space in nature. *Biosphere*, *8*(3), 258–267.

Zhizhin, G. V. (2016b). The structure, topological and functional dimension of biomolecules. *J. Chemoinformatics and Chemical Engineering*, *5*(3), 44–58.

Zhizhin G.V. (2017). Dimensions of compounds in supramolecular chemistry. *International Journal Chemical Modeling, 5*(2).

Zhizhin, G. V., & Bolshakova, N. N. (2000). Lone waves in the populations of unicellular animal organisms. *Mathematical Modelling*, *12*, 55–65.

Zhizhin, G. V., & Diudea, M. V. (2016). Space of nanoworld. In M. V. Putz & C. M. Marius (Eds.), *Sustainable nanosystems, development, properties, and applications* (pp. 214–235). New York: IGI Global.

Zhizhin, G. V., Khalaj, Z., & Diudea, M. V. (2016). Geometrical and topology dimensions of the diamond. In A. R. Ashrafi & M. V. Diudea (Eds.), *Distance, symmetry and topology in carbon nanomaterials* (pp. 167–187). New York: Springer. doi:10.1007/978-3-319-31584-3_12

Zhizhin, G. V., & Larina, T. I. (1994). Standing waves gas chemical reactions in porous inert media. *Combust*, *4*, 11–20.

Zhizhin, G. V., & Obukhova, I. A. (1995). Plane model of forced (non-adiabatic) wave of exothermic reaction in the condensed medium. *Mathematical Modelling*, *10*(7), 47–58.

Zhizhin, G. V., & Obukhova, I. A. (1997). Gel-effect in radical polymerization. *Mathematical Modelling*, *9*(11), 3–13.

Zhizhin, G. V., & Onattsky, P. L. (1981). The motion of particles in an unsteady flow of an incompressible liquid laminate. *Proceedings of the Academy of Sciences of the USSR. Mechanics of Fluid and Gas, 4,* 1399 - 1402.

Zhizhin, G. V., & Poritskaya, I. J. (1994). Self-regulating chemical reactions of n-th order in condensed matter. *Combust*, *6*, 61–68.

Zhizhin, G. V., & Segal, A. S. (1985). Statistical model of the three-phase flow reactor mixing phases with different temperatures. *Theoretical Foundations of Chemical Engineering*, *3*, 347–353.

Zhizhin, G. V., & Segal, A. S. (1986). Stationary flow in channels when the autowave propagation of chemical reaction with the dramatic increase in viscosity. *Journal of Applied Mechanics and Technical Physics*, *1*, 61–68.

Zhizhin, G. V., & Segal, A. S. (1988). Hydrodynamic stability spherical reaction front, accompanied by a strong increase in viscosity. *Proceedings of the Academy of Sciences of the USSR. Mechanics of Fluid and Gas, 3,* 46 - 53.

Zhizhin, G. V., & Ufimtsev, A. A. (1977). On the flow in the plane laminar boundary layer dilatant fluids. *Proceedings of the Academy of Sciences of the USSR. Mechanics of Fluid and Gas, 5.* 164 - 8.

Zhizhin, G. V., & Ufimtsev, A. A. (1978). The currents in the flat laminar boundary layer pseudoplastic fluids. *Proceedings of the Academy of Sciences of the USSR. Mechanics of Fluid and Gas, 1,* 186 - 8.

KEY TERMS AND DEFINITIONS

Branching of the Chain of the *D*-Glucose Molecule: The formation of branches in a chain of carbon atoms in the molecule of α-*D*-glucose. Such branches in a chain of the carbon atoms of the molecule β-*D*-glucose are impossible. This follows from the representation of glucose molecules in the form of a polytope of higher dimension.

Enantiomorphism (Chirality) of Biomolecules: The possibility of changing the mutual arrangement of hydrogen atoms and hydroxyl groups in biomolecules (polytopes of higher dimension), leading to a change in their properties.

Hybridization of Electronic Orbitals: This is the interaction of the electronic orbitals of atoms entering the molecule, leading under certain conditions to the formation of higher dimensionality of molecules.

Spiral Peptide Chain: The formation of a spiral chain of protein molecules, as a consequence of the higher dimension of protein molecules.

The First Coordination Sphere of Fe-Porphyrin: A set of nitrogen atoms bound by a covalent bond to an iron atom. The dimension of the coordination sphere upon addition of the oxygen atom increases from 5 to 6.

The Functional (Topological) Dimension of a Molecule: The dimension of a convex polytope, as a model of a molecule, at the vertices of which not only individual atoms but also functional groups of the molecule can be located.

Chapter 4
Convex Semi-Regular Polytopes

ABSTRACT

The geometry of polytopes of higher dimension having deviations from the conditions for the correctness of the geometric figure is considered. These deviations reflect the shapes of the molecules of the chemical compounds studied in Chapters 1-3. From the validity conditions in all cases the condition of topological equivalence of the vertices of the polytope is preserved. All these polytopes are called semi-regular. We study the hierarchical filling of spaces with polytopes of higher dimension, different from the well-known filling of spaces with spheres of constant diameter. The considered fillings characterize the distribution of atoms in nanostructures, in which the growth centers are distributed throughout the volume of the structure.

GOLDEN HYPER-RHOMBOHEDRON: TRANSLATIONAL BASIS OF QUASI-CRYSTALS

Finding in 1982 of ordered structures deprived (as it seemed) of translational symmetry (Shechtman et al., 1984), next called "quasicrystals", had marked the beginning of numerous cycles of papers and books devoted to the experimental and theoretical study of these unusual materials. But later it was found that the diffraction patterns of quasicrystals have a latent periodicity (Zhizhin, 2014), if we consider the diffraction pattern as a projection of a structure from a space of higher dimension. Figure 1 shows a typical diffraction pattern of intermetallic compounds.

DOI: 10.4018/978-1-5225-4108-0.ch004

Figure 1. Electron diffraction pattern of compound $Al_{72}Ni_{20}Co_8$ (Eiji Abe et al., 2004)

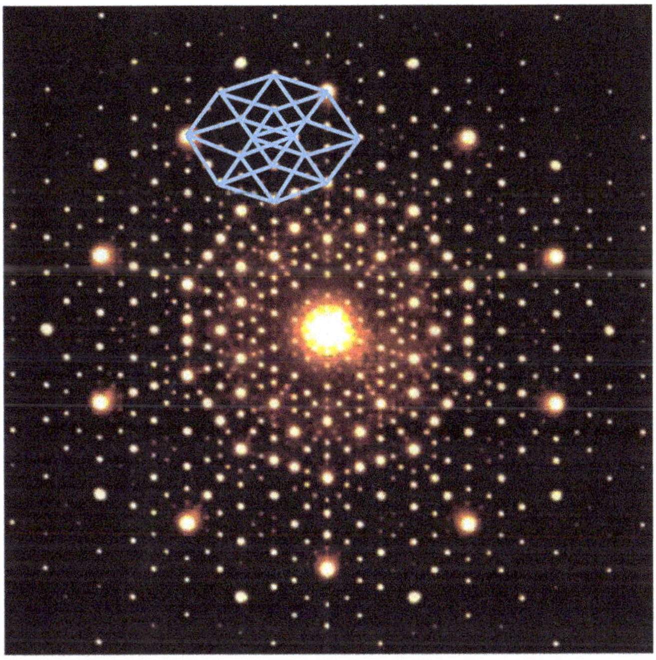

Similar structures have other intermetallic compounds involving transition metals, for example: Al_6Mn (Shechtman et al., 1984), $Al_{70}Fe_{20}W_{10}$ (Mukhopadhyay et al., 1993), $Ti_{54}Zr_{26}Ni_{20}$ (Zhang & Kelton, 1993).

Figure 1 shows that the luminous points, which are a reflection of light from the impact of the electron beam, form five families of parallel lines oriented with respect to each other at angles of a multiple of 72 degrees. The distances between the parallel lines and the angles are determined by the golden section. A geometrical model of the structure of the diffraction patterns of quasicrystals was constructed (Shevchenko, Zhizhin & Mackey, 2013a; Shevchenko, Zhizhin & Mackey, 2013b; Zhizhin, 2014; Zhizhin & Diudea, 2016). It was shown that the elementary cell of this geometric structure is a polytope of dimension 4, which was called a gold hyper-rhombohedron. This cell is plotted on the diffraction pattern in Figure 1 with solid segments of light lines. I can see that it passes through the luminous points of the diffraction pattern observing its geometry. This cell fills the entire space with the translation reflected by the diffraction pattern. To determine the regularities of this cell, it is depicted in Figure 2 on an enlarged scale.

Figure 2. Golden hyper-rhombohedron

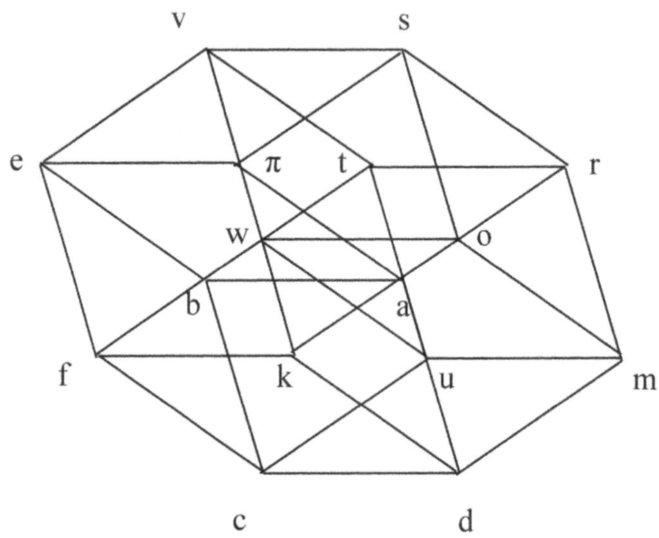

The polytope in Figure 2 consists of 8 three-dimensional polyhedrons whose faces are determined by construction with a golden section. Namely, all two-dimensional faces represent the same diamonds with angles $\varphi = \arccos\frac{1}{2\tau}, \gamma = \pi - \varphi$, where τ is golden section. The three-dimensional faces (rhombohedra) are adjacent to each other along the flat faces, so that all planar faces are common to some two rhombohedrons, which is necessary for the polytope to exist. The polytope in Figure 2 has the same number of vertices (16), edges (32), planar (24) and three-dimensional (8) faces as 4 - cub. Therefore, substituting these quantities of elements of different dimensions into the Euler-Poincaré equation (2) in Chapter 1, we get that this equation is satisfied for a value of dimension equal to 4. The difference from the 4-cube is that its flat faces are rhombs, and not squares. Therefore, the golden hyper-rhombohedron is a semi-regular polytope, since its two-dimensional faces are not regular polygons, although all vertices are compatible with motion. The periodic part of the structure of quasi-crystals is the translational filling of a four-dimensional space by a golden hyper-rhombohedron. The edges emanating from its vertices carry out the product of the rhombohedron into a segment. Just like a 4-cube is obtained by multiplying the cube by an edge. As a result, the number of elements of different dimensions in the gold hyper-rhombohedron is the same as in the 4-cube. The difference from the 4-cube lies in the fact that all its metric relationships are based on the golden section.

It was shown (Zhizhin, 2014, Zhizhin& Diudea, 2016) that the lattice of the vertices of the golden hyper-rhombohedron during its translation highlights the latent periodicity of the diffraction patterns. On this lattice can be built all the Platonic bodies, the Bravais (Bravais, 1848) and Delone (Delone, Padurov& Alecsandrov, 1934; Delone, 1937) cells, Voronoi (Voronoi, 1908) cells. Figure 3 representing the lattice of golden hyper - rhombohedron vertices, as an example, shows how from the vertices of the golden hyper-rhombohedron one can obtain projections of Plato's bodies: tetrahedron, cube, octahedron, icosahedron, dodecahedron.

Moreover, these figures can vary in size. It is essential that in addition to the already built golden hyper-rhombohedron of certain size, these polytopes of other dimensions can be built on vertices lattice (see Figure 4).

On the Figure 3 and Figure 4 the vertices of golden hyper-rhombohedron big size is indicated of a white circles. On Figure 4 besides golden hyper -rhombohedron little size is indicated of a black segments.

The physical nature of the process of scaling in phase transitions of the second kind and in quasicrystals is a change in the structure during the phase transition of the second kind (Kadanoff, 1966), and in the processes of formation of quasi-crystals during annealing. However, if Kadanoff assumed for phase transitions the merging of smaller cells to larger ones, here the analysis of golden hyper-rhombohedron vertices lattice shows that the process of scale

Figure 3. Plato`s bodies on the lattice of golden hyper-rhombohedron

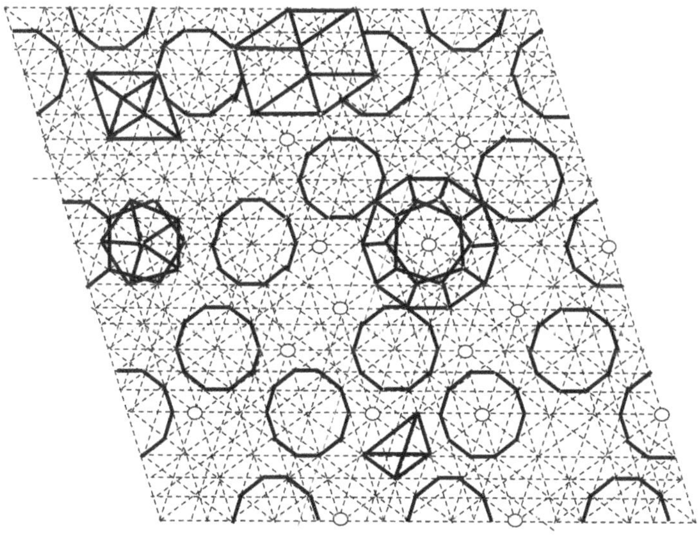

Figure 4. Golden hyper-rhombohedrons of different scale on the lattice of golden hyper-rhombohedron vertices

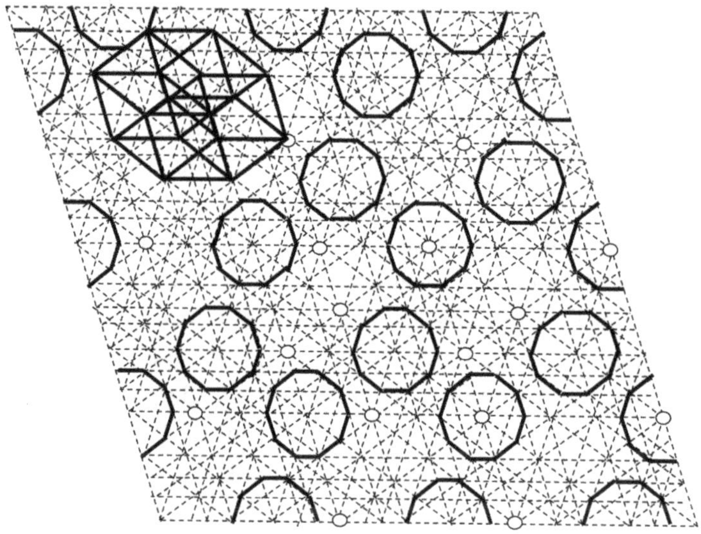

changing occurs hierarchically, i.e. a smaller hyper-rhombohedron cell is not combined with similar ones but simply increases in hierarchical mode, absorbing more and more of lattice sites. On the diffraction pattern (Figure 4), as an example, the projection of golden hyper-rhombohedron with the edge of a certain size is applied. It's clear that hyper-rhombohedrons of other sizes can be distinguished on the diffraction pattern. The analysis of diffraction pattern geometry shows that the similarity coefficient of hyper-rhombohedrons is the golden section τ. Raising τ in positive and negative integer powers one can receive hyper-rhombohedrons of appropriate sizes, which can be identified on the diffraction pattern in the range of its solvability. It is essential that the coordinates of the vertices of the golden hyper-rhombohedron are expressed in forth-dimensional space in whole numbers. Let's choose the top *a* in Figure 2 as the beginning of coordinates system *(x, y, z, h)*. Let's direct the axis *x* along the edge *ar,* the axis *y* along the edge *aπ*, the axis *z* along the edge *ad,* the axis *t* along the edge *ab*. Let's assume that the length of hyper-rhombedron edge is a unit. Then the coordinates of the vertices of the hyper-rhombohedron *aπebdkfc* are expressed in whole numbers *a(0, 0, 0, 0), π(0, 1, 0, 0), e(0, 1, 0, 1), b(0, 0, 0, 1), k(0, 1, 1, 0), f(0, 1, 1, 1), c(0, 0, 1, 1), d(0, 0, 1, 0).* Now let's to change the coordinate x from 0 to 1. Then the coordinates of a*πebdkfc* hyper-rhombohedron vertices also to change v*(1, 1, 0, 1), t(0, 0, 0, 1),w(1, 1, 1, 1), u(1, 0, 1, 1), s(1, 1, 0, 0), r(1, 0, 0, 0), o(1, 1,*

1, 0), m(1, 0, 1, 0). It's obvious that at hyper-rhombohedron translation the coordinates of all lattice vertices formed by hyper-rhombohedron vertices are integers. Filling the space of hyper-rhombohedrons of other sizes also forms integral lattice, if the size of hyper-rhombohedron edges is mistaken for one. It's interesting that on the lattice of golden hyper-rhombohedron vertices in the space 4D it is possible to construct the projections of the golden rhombohedron from a space of even more dimension. Figure 5 shows a projection of golden rhombohedron from the space 5D on the plane 2D.

It is clear that this projection can be laid on the lattice of golden hyper-rhombohedron vertices in the space 4*D* in Figure 3. The 5 – rhombohedron is the product of hyper-rhombohedron by one-dimension segment. If length of the segment equal of length edges of the hyper-rhombohedron, so the 5-rhombohedron is semi-regular polytope too. The 5 - rhombohedron can also fill the space of quasi-crystal displayed by the diffraction pattern in Figure 1, i. e. one can say that the quasi-crystal has a subspace of dimension 5. Similarly, a rhombohedron of even more dimension can be built on the diffraction pattern. This leads to the conclusion that quasi-crystals have subspaces of any integer dimension.

From the said above it follows that the lattice of the vertices of golden hyper-rhombohedrons have the universal nature.

Figure 5. The golden rhombohedron in space 5D

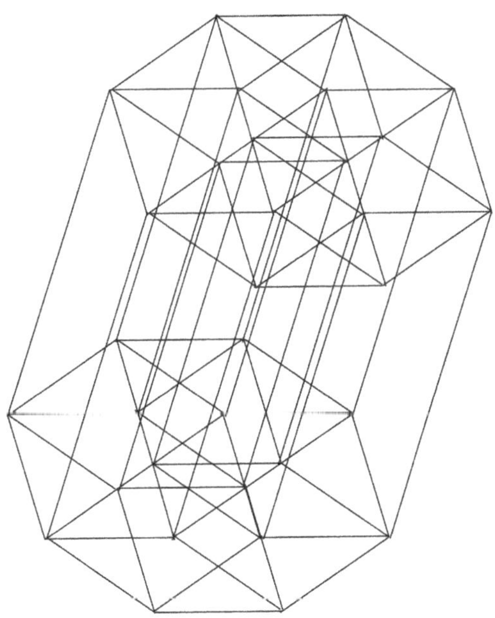

It is also important to note that on the diffraction patterns of intermetallides (Figure 1) one can see not only the latent periodicity of the structure in spaces of higher dimension, but also the hierarchical filling of spaces in the vicinity of almost any point of the structure. It is expressed in an increase in the size of the figures around an arbitrary point of the structure while maintaining the shape of the figure.

GEOMETRICAL CHARACTERISTICS OF THE CONVEX SEMI-REGULAR POLYTOPES

Investigations of the structures of molecules of chemical compounds of elements of the periodic system in Chapters 1-3 showed that in rare cases the molecules have the appearance of a regular convex polytope. Usually different deviations from the condition for the regularity of the polytope are encountered. These deviations arise either in metric or topological terms. Therefore, to continue the study of the structure of molecules of chemical compounds in a space of higher dimensions it is necessary to consider convex polytopes with deviations from the condition of their regularity: the compatibility of all vertices of a polytope with motion and regularity, as well as the equality of all faces of a polytope of a certain dimension. We shall call a convex polytope semi-regular if all its vertices are compatible with the motion, but the remaining validity conditions are violated in one way or another. At present, semi-regular polytopes have been studied very little. For the first time, there was announced the existence of convex semi-regular polytopes in four-dimensional space in Gosset's work (Gosset, 1900). It was a short note, containing only a statement about the existence of three four-dimensional semi-regular finite polytopes with an indication of their composition. Elte (Elte, 1912) came to the same results. Only in 1988 the validity of Gosset's statements was publicly mathematically confirmed (Makarov, 1988). In all the works mentioned, polytopes were considered in which all two-dimensional faces are the same - regular triangles, but the three-dimensional faces in the same polytope can be different. In 1900 and 1910 Stott's work (Stott, 1900; Stott, 1910) was published, in which, regardless of Gosset's work, semi-regular polytopes were considered, and the presence of a set of two-dimensional faces in the same polytope of different regular polygons was allowed. It is significant that in all these works no images of some semi-regular polytopes were shown. Images of semi-regular polytopes appeared in print only after 2013 in the works of Zhizhin (Zhizhin, 2013, Zhizhin, 2014, Zhizhin & Dudea, 2016).

The golden hyper-rhombohedron considered in the previous section represents an example of a semi-regular convex polytope in which plane faces are not regular polygons, but all vertices of the polytope are indent. In Zhizhin's paper (Zhizhin, 2014), another type of convex polytopes was discovered, in which edges with different incidence values for three-dimensional facets can be simultaneously present. Moreover, all the vertices of such polytopes are indent, so such polytopes can also be called semi-regular. Topologically, the compatibility condition for vertices in motion can be weakened, considering the vertices of polytopes to be indent, from which the same number of edges emanate. Such polytopes can also be considered semi-regular if some other validity conditions are violated.

CONVEX SEMI-REGULAR POLYTOPES WITH UNIFORM EDGES

In the book "World 4D" (Zhizhin, 2014) was proofed the theorem: From each convex three-dimensional polyhedron in the set of regular and semi-regular polyhedrons, one can go by means of a finite number of geometric transformations to any other polyhedron from this set of polyhedrons. Two geometric operations are used in the proof of the theorem. The operation t, consisting in cutting off the polyhedral angles of a regular polyhedron with their replacement by regular polygons with the same lengths of the sides as the edges of the remaining parts of the polyhedron. The operation t^2 consisting in repeating the operation t of cutting out the polyhedral angles of the obtained semi-regular polyhedron after performing the operation t. In this case, the necessary deformation of the dihedral angles and the lengths of the sides of the polygons is used in order to obtain closed bodies with the same lengths of sides in the whole body.

The operations t and t^2 can to use for obtaining the semi-regular polytopes of higher dimension. The semi-regular polytopes in this case will be have hyper-faces of the regular and semi-regular polytopes of the different form. The edges of the semi-regular polytopes will be uniform, i.e. it have equal meaning incidence of the tree-dimension polytopes throughout polytope of the higher dimension.

For notation of the semi-regular polytopes will be used the products of notation operation by notation of the regular polytopes of the symbol Schläfli (Schläfli, 1855). For the four-dimension polytopes the symbol Schläfli has look $\{i_1, i_2, i_3\}$, where i_1 is the number of vertices incident one flat face of the polytope, i_2 is the number of the flat faces of the polytope incident one vertex of the polytope, i_3 is the number tree-dimension faces of the polytope incident one the edge.

Obtaining semi-regular polytopes is reasonable to begin with the analysis of the simplest regular polytope of a four-dimensional regular simplex $\{3,3,3\}$. Applying the clipping operation *t*, i.e. replacing the edges by one third of the edge in its middle, we obtain a semi-regular polytope $t\{3,3,3\}$ whose set of hyper-faces consists of 5 tetrahedrons and 5 truncated tetrahedrons (Figure 6.)

Applying the clipping operation *t* to hyper-cube $\{4,3,3\}$, we obtain a semi-regular polytope $t\{4,3,3\}$ whose set of hyper-faces consists of 16 tetrahedrons and 8 truncated cube (Figure 7).

Applying the clipping operation *t* to 4-cross-polytope $\{3,3,4\}$, we obtain a semi-regular polytope $t\{3,3,4\}$ whose set of hyper-faces consists of 8 octahedrons and 16 truncated tetrahedrons (Figure 8).

Applying the clipping operation t^2 to regular simplex $\{3,3,3\}$, we obtain a semi-regular polytope $t^2\{3,3,3\}$ whose set of hyper-faces consists of 5 octahedrons and 5 tetrahedrons (Figure 9).

In view of the fact that in the projection in Figure 9 a part of the edges overlap, it is difficult to trace the existence of 5 octahedrons and 5 tetrahedrons in the figure. In order to see them enough to slightly to dismiss the coincidence vertices. This is done in Figure 10.

Then the octahedrons are figures 4, 6, 1, 10, 5, 8; 4, 10, 7, 9, 5, 3;10, 7, 2, 8, 1, 3; 4, 9, 2, 8, 3, 6; 5, 6, 2, 7, 1, 9. There are tetrahedrons 1, 5, 7, 10; 1, 6, 2, 8; 9, 2, 7, 3; 10, 8, 3, 4; 5, 4, 9, 6.

Applying the clipping operation t^2 to hyper-cube $\{4,3,3\}$, we obtain a semi-regular polytope $t^2\{4,3,3\}$ whose set of hyper-faces consists of 16 tetrahedrons and 8 cube-octahedrons (Figure 11).

If we denote *O* - octahedron, *T* - tetrahedron, *I* - icosahedron, *D* - dodecahedron, *C* - cube, *ID* – icosadodecahedron, *CO* - cuboctahedron, *tT* – truncated tetrahedron, *tC* – truncated cube, *tO* – truncated octahedron, *tD*

Figure 6. The four-dimensional semi-regular polytope $t\{3,3,3\}$

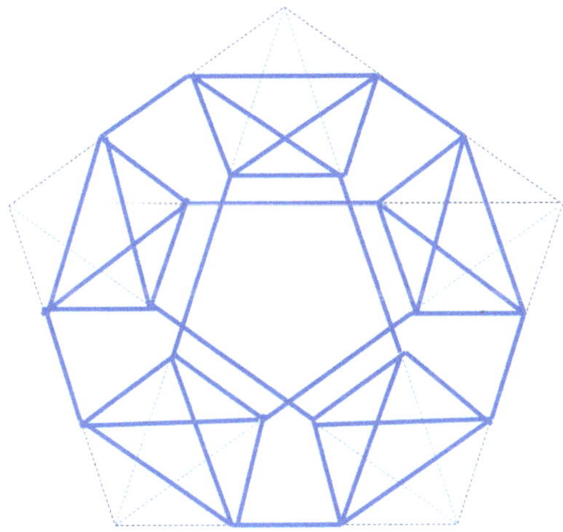

Figure 7. The four-dimensional semi-regular polytope $t\{4,3,3\}$

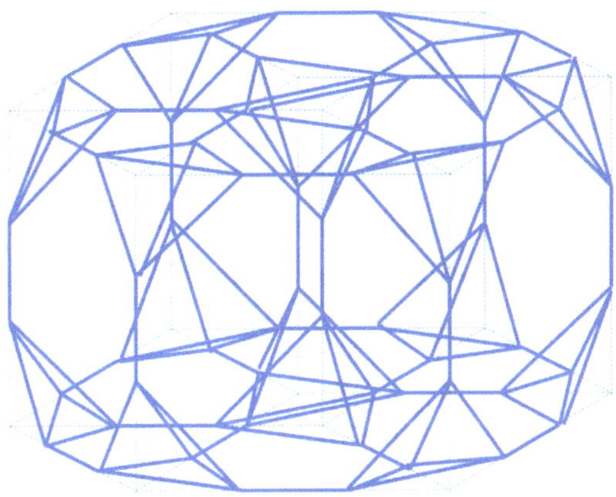

– truncated dodecahedron, *tI* – truncated icosahedron, then it is possible to enumerate all the obtained convex semi - regular four-dimensional polytopes. Some of these polytopes are not represented in the drawings because of their cumbersomeness. A list of all semi-regular polytopes obtained is given in Table 1, indicating the composition of the set of hyper-faces (the number of definite types of the hyper - faces is indicated in parentheses).

Figure 8. The four-dimensional semi-regular polytope $t\{3,3,4\}$

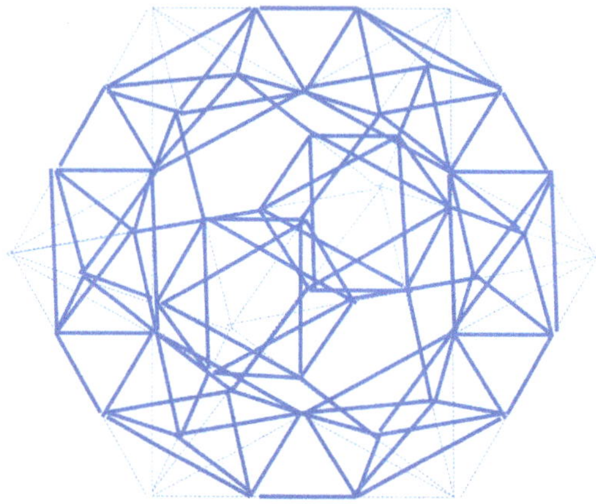

Figure 9. The four-dimensional semi-regular polytope $t^2\{3,3,3\}$

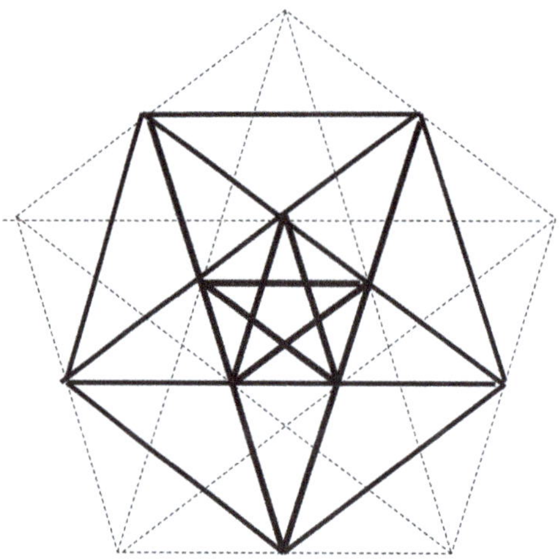

In Table 1 the first three semi-regular polytopes were established by Gosset (Gosset, 1900). In addition to the operations t and t^2 for obtaining semi-regular polytopes, the known (Coxeter, 1963) transformation of an octahedron into an icosahedron was used. Obviously, by known projections on a two-dimensional plane of regular polytopes of dimension greater than four, applying the introduced operations one can obtain semi-regular polytopes of dimension greater than four.

Figure 10. The four-dimensional semi-regular polytope $t^2\{3,3,3\}$ *with dismissal vertices*

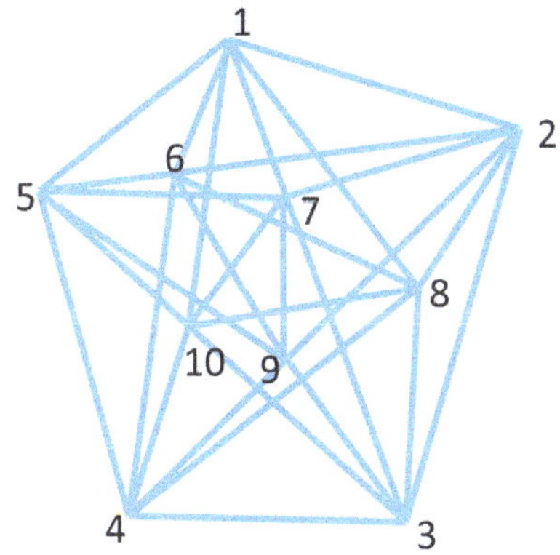

Figure 11. The four-dimensional semi-regular polytope $t^2\{4,3,3\}$

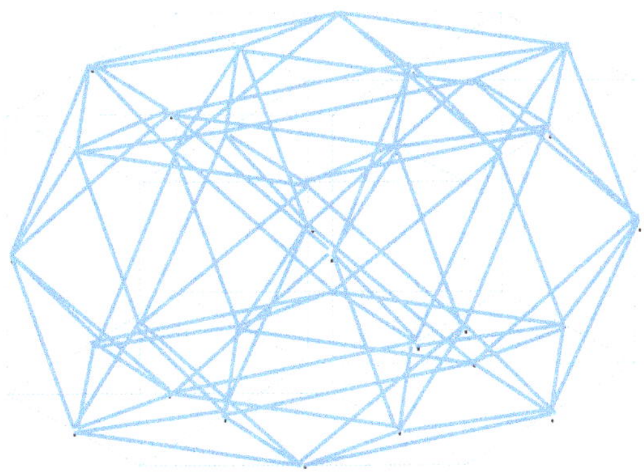

Table 1. The convex semi-regular four-dimension polytopes

N	Types and Number of the Hyper-Faces Semi-Regular Polytopes
1	$O(5)$, $T(5)$
2	$I(24)$, $T(120)$
3	$I(120)$, $O(600)$
4	$tT(5)$, $T(5)$
5	$tC(8)$, $T(16)$
6	$tT(16)$, $O(8)$
7	$tO(24)$, $C(24)$
8	$tD(120)$, $T(600)$
9	$tT(600)$, $I(120)$
10	$CO(8)$, $T(16)$
11	$tI(120)$, $tT(600)$
12	$CO(24)$, $C(24)$
13	$ID(120)$, $T(600)$
14	$I(10)$, $T(85)$

CONVEX SEMI-REGULAR POLYTOPES WITH DIFFERENT INCIDENCE VALUES OF EDGES TO THREE-DIMENSIONAL FACES

Structural chemistry is of interest for polytopes of higher dimension, the faces of which are regular figures, but there are deviations from homogeneity in the properties of polytope elements of different dimensions. In the general case, such higher-dimensional polytopes can be called regular-faceted polytopes of higher dimension. Three-dimensional regular-faceted polyhedrons were investigated in the works of Zalgaller (Zalgaller, 1967) and Zhizhin (Zhizhin, 2009).

As a proof of the existence of regular-faceted convex polytopes of higher dimension, we consider, for example, the following 9 metrically equal regular tetrahedrons with the vertices indicated in brackets: 1(1, 2, 3, 4), 2(1, 2, 3, 5), 3(1, 2, 5, 6), 4(1, 2, 4, 6), 5(2, 3, 4, 6), 6(2, 3, 4, 5), 7(1, 4, 5, 6), 8(1, 3, 4, 5), 9(3, 4, 5, 6).

It is easy to verify that every triangular face of any of the nine tetrahedrons is common to two tetrahedrons from the list. There is sufficient to construct a graph in which the vertices are the tetrahedrons from the indicated list, and each edge of the graph will correspond to the common triangular face between two tetrahedrons corresponding to the vertices of the edge of the graph (Figure 12).

Figure 12. The graph of set nine listed tetrahedrons

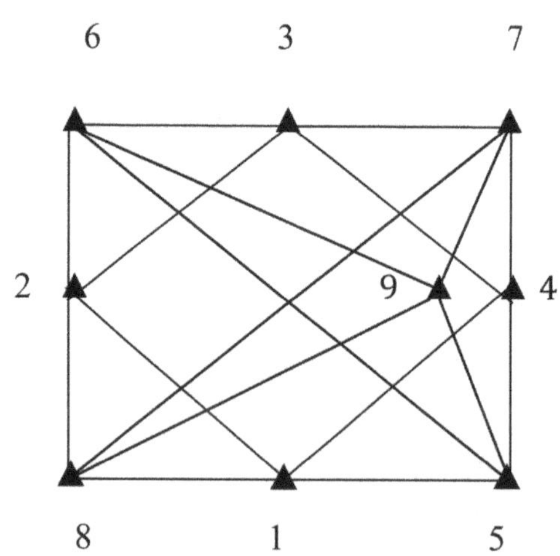

Table 2 lists the edges of the graph and the corresponding triangular faces of the nine tetrahedrons along which these tetrahedrons contact each other.

Since each face of nine tetrahedrons is closed by the face of some other tetrahedron of the indicated set of tetrahedrons, the figure made up of these tetrahedrons is closed and represents a certain polytope. Constructing the projection of this polytope onto a two-dimensional plane leads to the proof that each of the six vertices of nine tetrahedrons is connected by edges with the all other vertices. Therefore, the projection onto the two-dimensional plane of this polytope coincides with the projection of the five-dimensional regular simplex (Zhizhin, 2014) onto the two-dimensional plane (Figure 13).

It follows from Figure 13 and Table 2 that the constructed polytope has 6 vertices ($f_0 = 6$), 15 edges ($f_1 = 15$), 18 flat faces ($f_2 = 18$), 9 three-dimensional faces ($f_3 = 9$). Substituting these values $f_i, i = 1,2,3,$ into the Euler-Poincaré equation (2) of the first chapter, we find that the Euler-Poincaré equation holds for dimension $n = 4$

$$6 - 15 + 18 - 9 = 0.$$

This proves that the polytope constructed has dimension 4. Thus, removing from a 5-dimensional regular simplex of six tetrahedrons is not changed its projection onto a two-dimensional plane, but led to a decrease in the dimension of the polytope, while preserving its closure and convexity.

It follows from Figure 13 and Table 2 that each of the six edges of the polytope (out of 15) is incidentally to the three tetrahedrons of the polytope, and 9 edges of the polytope are incidentally to the four tetrahedrons of the

Table 2. Correspondence of the edges of the graph in Figure 12 to triangular faces of tetrahedrons

Edge of Graph	Face of Tetrahedron	Edge of Graph	Face of Tetrahedron	Edge of Graph	Face of Tetrahedron
1-2	123	1-5	243	1-4	124
1-8	143	2-8	135	2-3	125
2-6	235	3-6	625	3-4	126
3-7	156	4-5	246	4-7	146
6-9	356	6-5	263	5-9	364
7-8	154	7-9	564	8-9	345

Figure 13. Polytope with hyper- faces of tetrahedrons 1, 2, 3, 4, 5, 6, 7, 8, 9

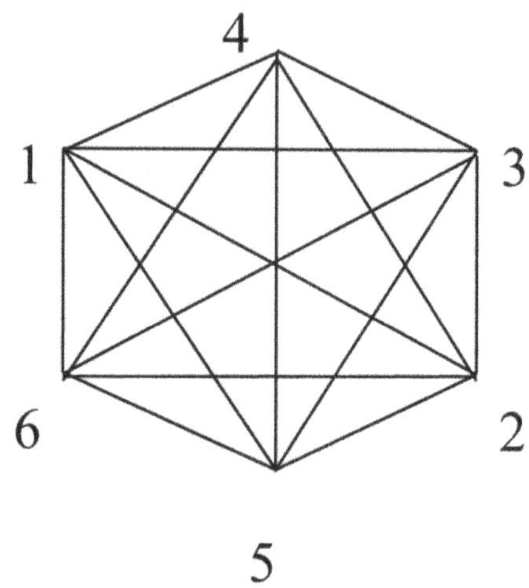

Table 3. Incidence of the edges polytopes on Figure 13 to tetrahedrons this polytopes

Edge of Polytope	Incident Tetrahedrons	Edge of Polytope	Incident Tetrahedrons	Edge of Polytope	Incident Tetrahedrons
25	2, 3, 6	16	3, 4, 7	24	1, 4, 5
13	1, 2, 8	45	7, 8, 9	36	5, 6, 9
12	1, 2, 3, 4	46	4, 5, 7, 9	35	2, 6, 8, 9
23	1, 2, 5, 6	14	1, 4, 7, 8	15	2, 3, 7, 8
56	3, 6, 7, 9	26	3, 4, 5, 6	34	1, 5, 8, 9

polytope (Table 3). Thus, the constructed convex polytope can be called polyincident.

We introduce 25 tetrahedrons with corresponding vertices: 1(1, 2, 3, 4), 2(1, 2, 3, 5), 3(1, 2, 5, 6), 4(1, 2, 6, 7), 5(1, 2, 4, 7), 6(2, 4, 7, 8), 7(4, 7, 8, 9), 8(4, 7, 9, 10), 9(4, 7, 1, 9), 10(2, 3, 4, 8), 11(3, 4, 8, 10), 12(3, 4, 9, 10), 13(3, 4, 1, 9), 14(2, 3, 5, 8), 15(3, 5, 8, 10), 16(3, 5, 9, 10), 17(3, 5, 1, 9), 18(2, 5, 6, 8), 19(5, 6, 8, 10), 20(5, 6, 9, 10), 21(1, 5, 6, 9), 22(2, 6, 7, 8), 23(6, 7, 8, 10), 24(6, 7, 9, 10), 25(1, 6, 7, 9).

It is easy to verify that every triangular face of any from the twenty-five tetrahedrons is common to two tetrahedrons from the list. There is sufficient to construct a graph in which the vertices are the tetrahedrons from the indicated list, and each edge of the graph will correspond to the common triangular face between two tetrahedrons corresponding to the vertices of the edge of the graph (Figure 14).

Table 4 lists the edges of the graph and the corresponding triangular faces of the twenty-five tetrahedrons along which these tetrahedrons contact each other.

Since each face of twenty-five tetrahedrons is closed by the face of some other tetrahedron of the indicated set of tetrahedrons, the figure made up of these tetrahedrons is closed and represents a certain polytope.

It follows from Figure 14 and Table 4 that the constructed polytope has 10 vertices ($f_0 = 10$), 35 edges ($f_1 = 35$), 50 flat faces ($f_2 = 50$), 25 three-dimensional faces ($f_3 = 25$). Substituting these values $f_1, i = 1, 2, 3,$ into the Euler-Poincaré equation (2) of the first chapter, we find that the Euler-Poincaré equation holds for dimension $n = 4$

$$10 - 35 + 50 - 25 = 0.$$

This proves that the polytope constructed has dimension 4.

It follows from Figure 14 and Table 4 that each of the ten edges of the polytope (out of 35) is incidentally to the five tetrahedrons of the polytope, and 25 edges of the polytope are incidentally to the four tetrahedrons of the polytope (Table 5). Thus, the constructed convex polytope can be called polyincident too.

Figure 14. The graph of set twenty-five listed tetrahedrons

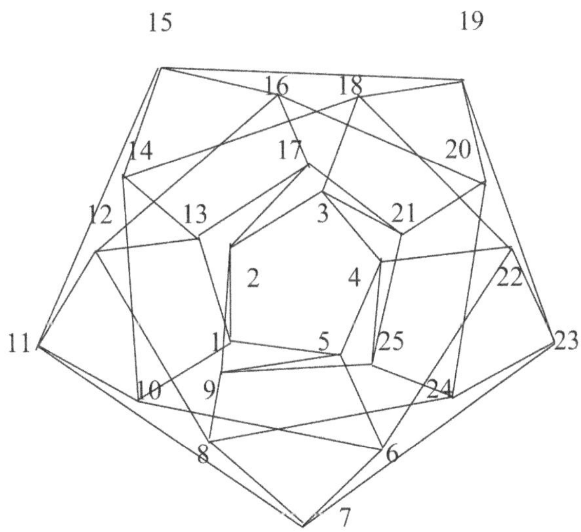

Table 4. Correspondence of the edges of the graph in Figure 14 to triangular faces of tetrahedrons

Edge of Graph	Face of Tetrahedron	Edge of Graph	Face of Tetrahedron	Edge of Graph	Face of Tetrahedron
1-2	1, 2, 3	1-5	1, 2, 4	1-10	3, 2, 4
1-13	1, 4, 3	2-3	1, 2, 5	2-14	3, 2, 5
2-17	1, 3, 5	3-21	6, 1, 5	3-4	1, 2, 6
3-18	2, 5, 6	4-5	2, 1, 7	4-22	2, 7, 6
4-25	1, 7, 6	6-5	2, 4, 7	5-9	1, 7, 4
6-7	7, 8, 4	6-10	2, 8, 4	6-22	2, 7, 8
7-8	4, 7, 10	7-11	4, 8, 10	7-23	7, 8, 10
8-9	4, 9, 7	8-12	4, 9, 10	8-24	7, 9, 10
9-13	1, 4, 9	9-25	1, 7, 9	10-14	2, 3, 8
10-11	3, 4, 8	11-12	3, 4, 10	11-15	3, 8, 10
12-13	3, 4, 9	12-16	3, 9, 10	13-17	1, 3, 9
14-15	3, 5, 8	14-18	2, 5, 8	15-16	3, 5, 10
15-19	5, 8, 10	16-17	3, 5, 9	16-20	5, 9, 10
17-21	1, 5, 9	18-19	5, 6, 8	18-22	2, 6, 8
19-20	5, 6, 10	19-23	6, 8, 10	20-21	5, 6, 9
20-24	6, 9, 10	21-25	1, 6, 9	22-23	6, 7, 8
23-24	6, 7, 10	24-25	6, 7, 9		

Table 5. Incidence of the edges polytopes with the graph on Figure 14 to tetrahedrons this polytopes

Edge of Polytope	Incident Tetrahedrons	Edge of Polytope	Incident Tetrahedrons	Edge of Polytope	Incident Tetrahedrons
2-3	1, 2, 6, 5, 7	1-4	1, 11, 16, 4, 21	1-5	2, 3, 12, 17, 22
2-6	3, 4, 8, 9, 10	5-7	5, 8, 13, 18, 22	7-8	14, 6,19, 9, 24
4-8	7, 10, 15,20, 25	6-9	11, 12, 13, 14,1 5	9-10	16, 17, 18, 19,20
10-3	21, 22, 23, 24,25	1-2	2, 3, 4, 1	1-3	1, 2, 21, 22
4-2	1, 7, 4, 10	4-3	1, 7, 21, 25	2-5	5, 2, 8, 3
3-5	2, 5, 22, 23	1-6	3, 4, 11, 12	5-6	3, 8, 12, 13
4-6	4, 10, 11, 15	2-7	5, 6, 8, 9	3-7	5, 6, 23, 24
2-8	6, 7, 9, 10	3-8	6, 7, 24, 25	6-7	8, 9, 13, 14
6-8	9, 10, 14, 15	1-9	11, 12, 16, 17	4-9	11, 15, 16, 20
5-9	12, 13, 17, 18	7-9	13, 14, 18, 19	8-9	14, 15, 19, 20
1-10	16, 17, 21, 22	4-10	16, 20, 21, 25	5-10	17, 18, 22, 23
7-10	18, 19, 23, 24	8-10	19, 20, 24, 25		

HIERARCHICAL FILLING OF SPACES WITH REGULAR AND SEMI-REGULAR POLYTOPES

The emergence of different levels of organization, i.e. the formation of a hierarchy of organization, is characteristic of both living and nonliving nature. The problem of hierarchical filling of spaces goes back to Kepler's work on hexagonal snowflakes (Kepleris, 1611). He considered the arrangement of balls of equal diameter in layers. In modern generalized crystallography,

Kepler's ideas develop in the works of Mackey (Mackey, 1981) and Kuo (Kuo, 2000). They consider layered filling of three-dimensional space with balls of equal diameter around some initial configuration. They assume that the centers of balls of equal diameter are located at the vertices of a polyhedron. Then the step of layer wise filling of the space around this polyhedron will be determined by the diameter of the ball, and not by the geometric properties of the polyhedron. We are also interested in the issue of transferring information about the source object only based on the characteristics of the object itself. From the electronic diffraction patterns of intermetallic compounds (Figure 1), it is seen that the bright points corresponding to the nodes of the structure diverge from the center with a consecutive increase in pitch. Such structures can not be answered by structures with layered filling of space with balls of equal diameter.

From each regular convex polygon in the plane, one can obtain a polygon of the same shape of a larger peak if through each vertex it draws lines perpendicular to the segments connecting the vertices with the center of the polygon. The intersection of these lines perpendicular to the segments forms a polygon similar to the original one. Moreover, the similarity coefficient is determined only by the geometric parameters of the initial polygon (Zhizhin, 2010).

From each regular convex polygon on the plane can obtain a larger polygon, similar to the original, in a different way. To do this, it is sufficient to continue to the intersection of the side of the polygon adjacent to both sides of each of the vertices of the polygon. By joining the intersection points, we obtain a new polygon similar to the original, with a similarity coefficient that depends only on the geometric parameters of the original polygon. Continuing the construction process, you can hierarchically fill the plane with regular polygons in both cases for any number of sides of the original polygon.

It is important that for such fillings of the plane the distances between the vertices increase in a geometric progression, as well as the distances between the nodes of the quasicrystal structure (see Figure 1). These two methods of hierarchical filling of the plane are transferred to a three-dimensional space (Zhizhin, 2012). It is shown that for the hierarchical filling of a three-dimensional space only the first method is applicable by regular polyhedron. In the case of a hierarchical filling of space by semi-regular polyhedron, only the second way of completing is applicable. Polytopes, the process of filling the space is accompanied by an alternation of polyhedron from a pair of dual. It is found that each polyhedron of a pair of duals forms a geometric progression with the same coefficient. The hierarchical filling of n-dimensional spaces by polytopes (Zhizhin, 2014) is considered. It is shown that when a n-dimensional space is filled with the regular simplex, it turns into itself and their sequence forms a geometric progression with coefficient $\sqrt{6}$ (the distance between the centers of two tetrahedrons adjacent to each other along an entire planar face is equal to $\frac{a}{\sqrt{6}}$, where a is the length of the edge of the tetrahedron).

In the process of hierarchical filling of n-dimensional space an n-cube becomes an n-cross-polytope, and the n-cross- polytope is transformed into an n-cube, as dual polytopes each other. Each polytope (n- cube and n –cross-polytope) formed its geometric progression with own coefficient (Zhizhin, 2014).

If the semi-regular polytope have regular hyper-faces, the process of hierarchical filling of space will be subject to the regularities of hierarchical filling of space with regular hyper-faces. If the hyper-faces of a semi-regular polytope are different from regular polyhedrons, the regularities of filling it with space will be based on the regularities of the hierarchical filling of space by semi-regular forms of smaller dimension.

Figure 15. The hierarchical filling of space with polytopes $t\{3,3,3\}$ *and* $\{3,3,3\}$

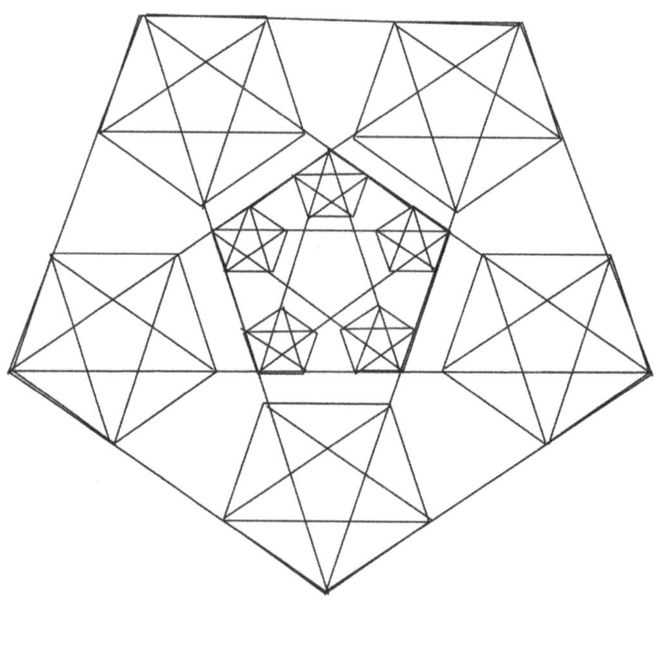

In view of the dramatic increase in the distance between the vertices of polytopes in the hierarchical filling of spaces and the sacking in nature to fill the voids, it is necessary to provide for the possibility of such filling in models of hierarchical filling of spaces by polytopes. This can be realized if the hierarchical filling of space does not take place from a single point in space, in principle, from any point in space. This is what we see in the diffraction patterns of intermetallides (Figure 1). It is also possible to combine in one structure the hierarchical filling of space with regular and semi-regular bodies simultaneously from different points in space.

Suppose that as the initial figure we have a semi-regular polytope $t\left\{3,3,3\right\}$ consisting of 5 truncated tetrahedrons and 5 tetrahedrons (Figure 6). The image on the plane of this polytope allows us to consider the process of attaching the regular simplex $\{3,3,3\}$ to it, including one tetrahedron of the polytope $t\left\{3,3,3\right\}$ in the composition of the regular simplex. Such a process can be continued and filled the entire space with a sufficiently dense distribution of vertices in it (Figure 15).

REFERENCES

Braveis, A. (1848). *Abhandlung über die Systeme von regelmässing auf einer Ebene oder im Raum vertheilten Puntsten*. Leipzig: Ostwald`s Klassiker.

Coxeter H.S.M. (1963). *Regular Polytopes*. New York: John Wiley & Sons, Inc.

Delone, B. (1937). Geometry of positive quadratic forms. *Uspekhi Mat. Nauk.*, *3*, 16–62.

Delone, B., Padurov, N., & Aleksandrov, A. (1934). *Mathematical foundations of the structural analysis of crystals*. Leningrad: ONTI.

Eiji, A., Yanfa, Y., & Pennycook, S. J. (2004). Quasicrystals as cluster aggregates. *Nature Materials*, *3*(11), 759–767. doi:10.1038/nmat1244 PMID:15516956

Elte, E. L. (1912). *The semi-regular polytopes of the hyperspace*. Groningen: University of Groningen.

Gosset, T. (1900). On the regular and semi-regular figures in space of *n* dimensions. *Messenger of Mathematics*, *29*, 43–48.

Kadanoff, L. P. (1966). Scaling Laws For Ising Models Near $\tau_c^* \tau_c^*$. *Physics, 2*, 263–272.

Kepleris J. (1611). *Strena, seu de nive sexangula*. Francofurti ad Moenum: apud Tampach.

Kuo, K. (2000). Mackay, Anti-Mackay, Double-Mackay, Pseudo-Mackay and Related Icosahedral Shell Clusters. *Structural Chemistry, 13*(3 / 4), 221–230.

Mackay, A. (1981). De nive quinquangula – About hexagonal snowflakes. *Crystallography, 26*(5), 910-919.

Makarov, M. V. (1988). On the derivation of four-dimensional semiregular polyhedrons. *Problems of Discrete Geometry. Mat. Research, 103*, 139–150.

Mukhopadhyay, N.K. (1993). Diffraction studies of icosahedral phases in $Al_{70}Fe_{20}W_{10}$. *Journal of Non-Crystalline Solids, 153-154*, 1193 – 1197.

Schläfli, L. (1855). Reduction d'une Integrale Multiple qui comprend l'are du cercle et l'are du triangle spherique comme cas particuliers. *Journal de Mathematiques, 20*, 359–394.

Shechtman, D., Blech, I., Gratias, D., & Cahn, J. W. (1984). Metallic phase with longerange orientational order and no translational symmetry. *Physical Review Letters, 53*(20), 1951–1953. doi:10.1103/PhysRevLett.53.1951

Shevchenko, V., Zhizhin, G., & Mackay, A. (2013a). On the structure of the quasicrystals in the high dimension space. In M. V. Diudea (Ed.), *Diamond and Related Nanostructures*. Dordrecht: Springer. doi:10.1007/978-94-007-6371-5_17

Shevchenko, V.Y., Zhizhin, G.V., & Mackay, A.L. (2013b). On the structure of quasicrystals in a space of higher dimension. *Izvestiya RAS. Chemical series, 2*, 269 - 274.

Stott A. B. (1900). On sertain series of sections of the regular four-dimensional hyper-solids. *Verhandelingen der Koninklijke Akademie van Wetenschappen te Amsterdam*, 3-7.

Stott A. B. (1910). Geometrical deduction of semiregular from regular polytopes and space fillings. *Verhandelingen der Koninklijke Akademie van Wetenschappen te Amsterdam,* 1 – 11.

Voronoi, G. (1908). Recherches sur les parallelloedres primitiss. *Grelle Journal*, *134*, 287–1908.

Zalgaller, V. A. (1967). Convex polyhedrons with regular faces. *Zap. Scientific. Sem. LOMI*. Moskow – Leningrad. *Science*, 5–221.

Zhang X. & Kelton K.F. (1993). High-order crystal approximant structures alloys $Ti_{54}Zr_{26}Ni_{20}$. *Journal of Non-Crystalline Solids*, *153-154*, 114 – 118.

Zhizhin G.V. (2009). *Convex right-faceted polyhedrons*. St. Petersburg: Polytehnika-service.

Zhizhin G.V. (2010). *Geometrical bases of the dissipative structures*. St. Petersburg: Polytechnika.

Zhizhin, G. V. (2012). Hierarchical filling of spaces with polytopes. In *Science and the progress of mankind. VII St. Petersburg meeting of Nobel Prize laureates*. St. Petersburg: Publishing House of Polytechnic University.

Zhizhin, G. V. (2013). Images of convex regular and semiregular n-dimensional polytopes. In *Proceedings of the 9th All-Russian Scientific School "Mathematical Research in the Natural Sciences"*. Apatity, Geological Institute KSC RAS.

Zhizhin, G. V. (2014). *World – 4D*. St. Petersburg: Polytechnic Service.

Zhizhin, G. V., & Diudea, M. V. (2016). Space of Nanoworld. In M. V. Putz & M. C. Mirica (Eds.), *Sustainable Nanosystems, Development, Properties, and Applications* (pp. 214–236). New York: IGI Global.

KEY TERMS AND DEFINITIONS

Golden Hyper-Rhombohedron: Semi-regular polytope of dimension *n*, which is an elementary cell of quasicrystals in *n*-dimensional space. In the four-dimensional space, it includes 8 three-dimensional rhombohedrons whose faces are determined by the golden section.

Latent Periodicity of Quasicrystals: The periodicity of quasicrystals in a space of higher dimension.

Scaling on the Lattice of the Vertices of the Golden Hyper-Rhombohedron: The lattice of the vertices of the golden hyper-rhombohedron allows a discrete scale change (asymptotic expansion from each lattice point).

Semi-Regular Convex Polytopes: Convex polytopes, in which some of the conditions for the correctness of polytopes are violated while other correctness conditions are preserved. Four-dimensional semi-regular convex polytopes include polytopes obtained by truncating regular convex polytopes. They have a lot of homogeneous edges, but facets are simultaneously different three-dimensional figures. Another example of semi-regular convex polytopes are polytopes, in each of it there are edges with different incidence of polyhedron facets.

Chapter 5
Polytopic Prismahedrons:
Fundamental Regions of the *n*-Dimension Nanostructures

ABSTRACT

The structure of polytopes—polytopic prismahedrons, which are products of polytopes of lower dimensionality—is investigated. The products of polytopes do not belong to the well-studied class of simplicial polytopes, and therefore their investigations are of independent interest. Analytical dependencies characterizing the structure of the product of polytopes are obtained as a function of the structures of polytope factors. Images of a number of specific polytopic prismahedrons are obtained, tables of structures of polytopic prismahedrons are compiled, depending on the types of polytopes of the factors. Polytopic prismahedrons can be considered as a result of the chemical interaction of molecules, which, from among which there is a polytope of a certain dimension.

THE STRUCTURE OF POLYTOPES AS A FUNCTION OF FACTORS STRUCTURE

While investigating diffraction patterns of quasi-crystals the golden hyper-rhombohedron with dimension 4 was built (Chapter 4). As it was noted, it may be isolated in the quasi-crystals diffraction patterns as a fundamental domain. It is formed as a product of the golden rhombohedron by a one-dimensional

DOI: 10.4018/978-1-5225-4108-0.ch005

segment. Pontryagin mentioned the structures resulting from the product of a polyhedron by a one-dimensional segment the cylinder ones (Pontryagin, 1976). In Chapter 4 it was noted that polytopes of dimension greater than 4 can be distinguished on the grid of vertices of a hyper-rhombohedron. We can say that the product of a polytope by a segment is a prism (Robertson, 1984) with a base in the form of a polytope. To distinguish it from a usual three-dimensional prism, we call it polytopic prism. Ziegler noted that the product of polytopes is not a simplex even if the factors are simplexes, so the polytopes are of considerable interest (Ziegler, 1995). Especially taking into account that a multi-dimensional world has its own peculiarities having no analogues in the three-dimensional world (Ziegler, 1995), contrary to some opposite statements (Panina, 2006). In this regard, the developed theory of simplicial polytopes (Fomenko, 1992; Pontryagin, 1976; Alexandrov, 1975) for the analysis of polytopes product becomes inapplicable, especially in the case of high-dimensional factors.

The product of two polytopes is the result of the product of one of them by one-dimensional edges of another polytope. Thus, the product of polytopes is a complex of polytopic prisms. Let's call this complex a polytopic prismahedron. The interest to the study of polytopic prismahedrons is connected, in the first place, with the novelty of this field and, secondly, with the fact that polytopic prismahedrons, due to their construction, can be, as we'll see later, "bricks" to fill the spaces of higher dimension face in face. The definition of polytopes product (Ziegler, 1995) does not give the possibility to specify the structure of the product as a function of the factors structures. There is the structure of product of polytopes having different structures of their factors is determined.

Theorem 1 (Zhizhin, 2015)

If we have convex polytopes of dimensions *n* and *m*, respectively denoted P^n and Q^m (or simply P and Q), then their product $P^n \times Q^m$, (or simply ×, when it is clear what polytopes are multiplied) has a face $F_{\times}^k$ with numbers

$$f_{\times}^k = \sum_{i=0}^{j} f_P^{k-i} f_Q^i , \tag{1}$$

$j = k,$, if $0 \le k < m; j = m,$ if $m \le k \le n + m; n \ge m$.

Here the symbol *f* indicates the number of faces, the superscript of *f* and *F* indicates the dimension of faces, the lower index indicates belonging of a face to the respective polytope.

Proof

To find the product of polytopes *P* and *Q* one has to find the product of geometric elements of different dimensions of one of the polytopes by geometric elements of different dimensions of another polytope. According to the definition of polytopes product (Ziegler, 1995), the product of vertices (elements of zero-dimension) of one of the polytopes by the vertices of another polytope is a set of vertices, the number of which is equal to the product of the number of vertices of one of the polytopes by the number of vertices of another polytope, i.e.

$$f_\times^0 = f_P^0 f_Q^0 .$$

The number of elements of dimension 1 of polytopes product consists of the product of number of elements of dimension 1 of one of polytopes by the number of elements of zero-dimension of another polytope and the product of the number of elements of zero-dimension of one of polytopes by the number of elements of the dimension 1 of another polytope, i.e.

$$f_\times^1 = f_P^1 f_Q^0 + f_P^0 f_Q^1 .$$

The number of elements of dimension 2 of polytopes product consists of the product of the number of elements of dimension 2 of one of the polytopes by the number of elements of zero dimension of another polytope, the product of the number of elements of dimension 1 of one of polytopes by the number of elements of dimension 1 of another polytope and the product of the number of elements of zero dimension of one of the polytopes by the number of elements of dimension 2 of another polytope, i.e.

$$f_\times^2 = f_P^2 f_Q^0 + f_P^1 f_Q^1 + f_P^0 f_Q^2 .$$

Continuing this process, we'll obtain the expression for the number of elements of dimension *k* in the product of polytopes *P* and *Q*

$$f_\times^k = f_P^k f_Q^0 + f_P^{k-1} f_Q^1 + \ldots + f_P^0 f_Q^k . \qquad (2)$$

The specific form of equation (2) depends on the ratio of k with the dimensions of polytopes P and Q. If k is greater than m, i.e. the dimension of the polytope Q, then in the series (2) several last items are lost because of the absence of faces of the polytope Q with dimension greater than m. If k is greater than n, i.e. the dimension of the polytope P, in the series (2) several first items are lost because of the absence of faces of the polytope P with the dimension greater than n.

If $k < m$ the series (2) can be written in the form

$$f_\times^k = \sum_{i=0}^{k} f_P^{k-i} f_Q^i . \qquad (3)$$

When $k \geq m$ summation by i in the series (3) is carried out only up to the value $i = m$

$$f_\times^k = \sum_{i=0}^{m} f_P^{k-i} f_Q^i . \qquad (4)$$

At that, in the row (4) the items corresponding to the values of $k - i > n$ are equal to zero. Equations (3) and (4) can be written as a single equation

$$f_\times^k = \sum_{i=0}^{j} f_P^{k-i} f_Q^i , \qquad (5)$$

where $j = k$, if $0 \leq k < m; j = m$, if $m \leq k \leq n + m; n \geq m$.

Theorem 1 is proved.

Theorem 1 determines the number of figures of different dimensions in the polytopes product, but it cannot identify in general terms what is the form of these figures. It follows from theorem 1 that the form of the figures of polytopes product depends on the form of the polytopes which are the factors of this product.

THE PRODUCTS OF POLYTOPES BY ONE–DIMENSIONAL SEGMENT

It is obvious that the product of a point a by a one-dimensional segment is the same segment with point on end coinciding with point a. The product of segment by segment is a quadrangle. The product triangle on the segment is a triangular prism with the bases equal to the triangle, and generatrices equal to the segment.

The Product of a Tetrahedron by a Segment

For the designation of a polytope let's agree to indicate not only its dimension but the number of its vertices by lower index of polytope symbol. Besides, we'll also indicate the number of its faces in brackets after the polytope sign, face dimension and the number of its vertices (for the faces with dimension greater or equal to 2). For example, a tetrahedron in this record has the form $P_4^3(4F_3^2)$. If a polytope is considered in this case as a face we'll denote it by the letter F. Let's call such a record a structural polytope formula. We need to get from structural formulas of polytopes as factors the structural formula of the polytopes product. Thus, in this case, we consider the product $P_4^3(4F_3^2)\times Q_2^1$. According to theorem 1 and the ratios (5) we have

$$f_\times^0 = f_P^0 f_Q^0 = 4 \cdot 2 = 8,$$

$$f_\times^1 = f_P^1 f_Q^0 + f_P^0 f_Q^1 = 6 \cdot 2 + 4 \cdot 1 = 16..$$

When calculating the number of faces of dimension greater than 1 we'll indicate in brackets after the sign of the faces number the designation of the corresponding face. This will give us an opportunity to get the composition of faces in the polytopes product. Thus,

$$f_\times^2 = f_P^2(F_{P3}^2) f_Q^0(F_{Q1}^0) + f_P^1(F_{P2}^1) f_Q^1(F_{Q2}^1) = 4 \cdot 2 + 6 \cdot 1 = 14. \quad (6)$$

For receiving of the composition of the faces in the dimension 2 polytopes product we multiply the faces in brackets in the product of the numbers of faces, observing the same rules as at the definition of the polytopes product, i.e. dimensions are added and the numbers of vertices are multiplied. At that, indexes P and Q are replaced by index ×. Then 14 faces obtained in (6), are deciphered so

$$14 = 8F^2_{\times 3} + 6F^2_{\times 4}. \qquad (7)$$

This means that 14 faces of dimension 2 include eight faces $F^2_{\times 3}$ (i.e., triangles) and six faces $F_{\times 4}$ (i.e., quadrangles). In future, equations similar to (6), (7) will be recorded in a single line.

For the faces of dimension 3, we have

$$f^3_\times = f^3_P(F^3_{P4})f^0_Q(F^0_{Q1}) + f^2_P(F^2_{P3})f^1_Q(F^1_{Q2}) = 1\cdot 2 + 4\cdot 1 = 6 = 2F^3_{\times 4} + 4F^3_{\times 6}.$$

Thus, 6 faces of dimension 3 consist of two tetrahedrons $P^3_4(4F^2_3)$ and four triangular prisms $P^3_6(3F^2_4, 2F^2_3)$.

For the faces of dimension 4, we have

$$f^4_\times = f^3_P(F^3_{P4})f^1_Q(F^1_{Q2}) = 1\cdot 1 = 1 = P^4_8.$$

According to the preceding equation we obtain the structural formula of the product

$$P^3_4(4F^2_3)\times Q^1_2 = P^4_8[4P^3_6(3F^2_4, 2F^2_3), 2F^3_4(4F^2_3)]. \qquad (8)$$

Therefore, the product of the tetrahedron by the segment is polytope 4D, which includes four triangular prisms and 2 tetrahedrons. You can call a polytope (8) tetrahedral prism. The picture of the polytope (8) is shown in Figure 1.

Figure 1. Tetrahedral prism

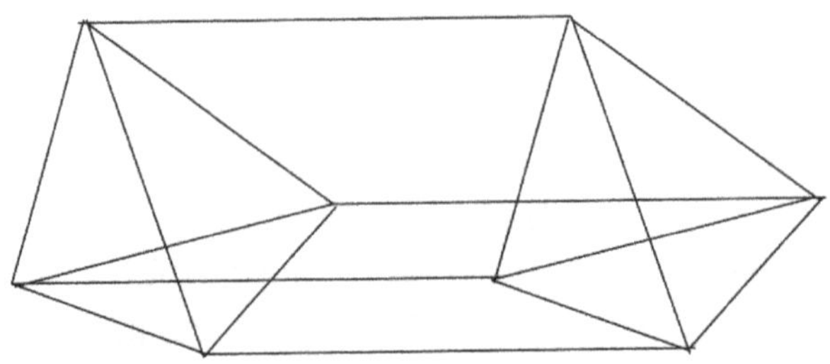

It is easy to verify that the Euler-Pioncare equation (2) in Chapter 1 for the polytope (8) is satisfied

$8 - 16 + 14 - 6 = 0.$

In future the verification of the Euler – Pioncare (Pioncare, 1895) equation satisfaction is omitted, as a reader can easily do it on his own.

The Product of a Cube by a Segment

The structural formula of a cube is $P_8^3(6F_4^2)$. It is necessary to find a structural formula of the product $P_8^3(6F_4^2) \times Q_2^1$. According to theorem 1 and formulas (5), we find

$$f_\times^0 = f_P^0 f_Q^0 = 8 \cdot 2 = 16,$$

$$f_\times^1 = f_P^1 f_Q^0 + f_P^0 f_Q^1 = 12 \cdot 2 + 8 \cdot 1 = 32,.$$

$$f_\times^2 = f_P^2(F_4^2) f_Q^0(F_{Q1}^0) + f_P^1(F_{P2}^1) f_Q^1(F_{Q2}^1) = 6 \cdot 2 + 12 \cdot 1 = 24 = 24F_{\times 4}^2,$$

$$f_\times^3 = f_P^3(F_{P8}^2) f_Q^0(F_{Q1}^0) + f_P^2(F_{P4}^2) f_Q^1(F_{Q2}^1) = 1 \cdot 2 + 6 \cdot 1 = 8 = 8F_{\times 8}^3,$$

$$f_\times^4 = f_P^3(F_{P8}^3) f_Q^1(F_{Q2}^1) = 1 \cdot 1 = 1 = P_{16}^4(8F_8^3). \tag{9}$$

Thus, we see that the product of a cube by a segment is $4D$ – cube (hypercube) containing 8 cubes.

The Product of an Octahedron by a Segment

The structural formula of an octahedron is $P_6^3(8F_4^2)$. It is necessary to find a structural formula of the product $P_6^3(8F_3^2) \times Q_2^1$. According to theorem 1 and formulas (5), we find

$$f_\times^0 = f_P^0 f_Q^0 = 6 \cdot 2 = 12,$$

$$f_\times^1 = f_P^1 f_Q^0 + f_P^0 f_Q^1 = 6 \cdot 1 + 12 \cdot 2 = 30,.$$

$$f_\times^2 = f_P^2(F_{P3}^2) f_Q^0(F_{Q1}^0) + f_P^1(F_{P2}^1) f_Q^1(F_{Q2}^1) = 8 \cdot 2 + 12 \cdot 1 = 28 = 16F_{\times 3}^2 + 12F_{\times 4}^2,$$

$$f_\times^3 = f_P^3(F_{P6}^2) f_Q^0(F_{Q1}^0) + f_P^2(F_{P3}^2) f_Q^1(F_{Q2}^1) = 8F_{\times 6}^3(3F_4^2, 2F_3^2) + 2F_{\times 6}^3(8F_3^2),$$

$$f_\times^4 = f_P^3(F_{P6}^3) f_Q^1(F_{Q2}^1) = 1 \cdot 1 = 1 = P_{12}^4.$$

According to the last equality we obtain the structural formula of the product

$$P_6^3(8F_3^2) \times Q_2^1 = P_{12}^4[8F_6^3(3F_4^2, 2F_3^2), 2F_6^3(8F_3^2)]. \tag{10}$$

Thus, the product of an octahedron by a segment is a polytope 4*D*, which includes 8 triangular prisms and two octahedrons. You can call the polytope (10) an octahedral prism.

The Product of an Icosahedron by a Segment

The structural formula of an icosahedron is $P_{12}^3(20F_3^2)$. It is necessary to find the structural formula of the product $P_{12}^3(20F_3^2) \times Q_2^1$. According to theorem 1 and formulas (5), we find $f_\times^0 = f_P^0 f_Q^0 = 12 \cdot 2 = 24$,

$$f_\times^1 = f_P^1 f_Q^0 + f_P^0 f_Q^1 = 30 \cdot 2 + 12 \cdot 1 = 72,$$

$$f_\times^2 = f_P^2(F_{P3}^2) f_Q^0(F_{Q1}^0) + f_P^1(F_{P2}^1) f_Q^1(F_{Q2}^1) = 20 \cdot 2 + 30 \cdot 1 = 70 = 40F_{\times 3}^2 + 30F_{\times 4}^2,$$

$$f_\times^3 = f_P^3(F_{P12}^2) f_Q^0(F_{Q1}^0) + f_P^2(F_{P3}^2) f_Q^1(F_{Q2}^1) =$$
$$1 \cdot 2 + 20 \cdot 1 = 22 = 20F_{\times 6}^3(3F_4^2, 2F_3^2) + 2F_{\times 12}^3,$$

$$f_\times^4 = f_P^3(F_{P12}^3) f_Q^1(F_{Q2}^1) = 1 \cdot 1 = 1 = P_{24}^4.$$

According to the last equality, we obtain the structural formula of the product

$$P^3_{12}(20F^2_3)\times Q^1_2 = P^4_{24}[20F^3_6(3F^2_4, 2F^2_3), 2F^3_{12}(20F^2_3)]. \quad (11)$$

Thus, the product of an icosahedron by a segment is a polytope $4D$, which includes 20 triangular prisms and 2 icosahedrons. You can call a polytope (11) icosahedral prism.

The Product of a Dodecahedron on a Segment

The structural formula of a dodecahedron is $P^3_{20}(12F^2_5)$. It is necessary to find the structural formula of product $P^3_{20}(12F^2_5)\times Q^1_2$. According to theorem 1 and formulas (5), we find

$$f^0_\times = f^0_P f^0_Q = 20 \cdot 2 = 40,$$

$$f^1_\times = f^1_P f^0_Q + f^0_P f^1_Q = 30 \cdot 2 + 20 \cdot 1 = 80,$$

$$f^2_\times = f^2_P(F^2_{P5}) f^0_Q(F^0_{Q1}) + f^1_P(F^1_{P2}) f^1_Q(F^1_{Q2}) = 12 \cdot 2 + 30 \cdot 1 = 54 = 24F^2_{\times 5} + 30F^2_{\times 4},$$

$$f^3_\times = f^3_P(F^2_{P20}) f^0_Q(F^0_{Q1}) + f^2_P(F^2_{P5}) f^1_Q(F^1_{Q2}) =$$
$$1 \cdot 2 + 12 \cdot 1 = 14 = 12F^3_{\times 10}(5F^2_4, 2F^2_5) + 2F^3_{\times 20}(12F^2_5),$$

$$f^4_\times = f^3_P(F^3_{P20}) f^1_Q(F^1_{Q2}) = 1 \cdot 1 = 1 = P^4_{40}.$$

According to the last equality, we obtain the structural formula of the product

$$P^3_{20}(12F^2_5)\times Q^1_2 = P^4_{40}[12F^3_{10}(5F^2_4, 2F^2_5), 2F^3_{20}(12F^2_5)]. \quad (12)$$

Thus, the product of a dodecahedron by a segment is a polytope $4D$ including 12 pentagonal prisms and 2 dodecahedrons. You can call the polytope (12) a dodecahedral prism.

The Product of an Egyptian Pyramid by a Segment

The structural formula of an Egyptian pyramid is $P_5^3(4F_3^2, \mathrm{F}_4^2)$. It is necessary to find the structural formula of product $P_5^3(4F_3^2, \mathrm{F}_4^2) \times Q_2^1$. According to theorem 1 and formulas (5), we find

$$f_\times^0 = f_P^0 f_Q^0 = 5 \cdot 2 = 10,$$

$$f_\times^1 = f_P^1 f_Q^0 + f_P^0 f_Q^1 = 5 \cdot 1 + 8 \cdot 2 = 21,$$

$$f_\times^2 = f_P^2(4F_{P3}^2, \mathrm{F}_{P4}^2) f_Q^0(F_{Q1}^0) + f_P^1(F_{P2}^1) f_Q^1(F_{Q2}^1) = \\ 5 \cdot 2 + 8 \cdot 1 = 18 = 8F_{\times 3}^2 + 10F_{\times 4}^2,$$

$$f_\times^3 = f_P^3(F_{P5}^2) f_Q^0(F_{Q1}^0) + f_P^2(F_{P3}^2, \mathrm{F}_{P4}^2) f_Q^1(F_{Q2}^1) \\ = 1 \cdot 2 + 5 \cdot 1 = 7 = 2F_{\times 5}^3 + 4F_{\times 6}^2 + F_{\times 8}^3,$$

$$f_\times^4 = f_P^3(F_{P5}^3) f_Q^1(F_{Q2}^1) = 1 \cdot 1 = 1 = P_{10}^4.$$

According to the last equality, we obtain the structural formula of the product

$$P_5^3(4F_3^2, \mathrm{F}_4^2) \times Q_2^1 = P_{10}^4[4F_6^3(3F_4^2, 2F_3^2), F_8^3(6F_4^2), 2\,\mathrm{F}_5^3(4\,\mathrm{F}_3^2, \mathrm{F}_4^2)].$$

Thus, the product of an Egyptian pyramid by segment is a polytope 4*D* including four triangular prisms, 1 quadrangular prism and 2 Egyptian pyramids. You can call the polytope a pyramidal prism.

The Product of a 4*D* - Simplex by a Segment

The structural formula of a 4*D*-simplex is $P_5^3[5F_4^3(4F_3^2)]$. It is necessary to find the structural formula of the product $P_5^3[5F_4^2(4F_3^2)] \times Q_2^1$. According to theorem 1 and formulas (5), we find $f_\times^0 = f_P^0 f_Q^0 = 5 \cdot 2 = 10$,

$$f_\times^1 = f_P^1 f_Q^0 + f_P^0 f_Q^1 = 10 \cdot 2 + 5 \cdot 1 = 25,$$

$$f_{\times}^{2}=f_{P}^{2}(F_{P3}^{2})f_{Q}^{0}(F_{Q1}^{0})+f_{P}^{1}(F_{P2}^{1})f_{Q}^{1}(F_{Q2}^{1})=10\cdot 2+10\cdot 1=30=20F_{\times 3}^{2}+10F_{\times 4}^{2},$$

$$f_{\times}^{3}=f_{P}^{3}(F_{P4}^{2})f_{Q}^{0}(F_{Q1}^{0})+f_{P}^{2}(F_{P3}^{2})f_{Q}^{1}(F_{Q2}^{1})=$$
$$5\cdot 2+10\cdot 1=20=10F_{\times 4}^{3}(4F_{3}^{2})+10F_{\times 6}^{2}(3F_{4}^{2},2F_{3}^{2}),$$

$$f_{\times}^{4}=f_{P}^{4}[5F_{P4}^{3}(4F_{3}^{3})]f_{Q}^{0}(F_{Q1}^{0})+f_{P}^{2}(F_{P4}^{2})f_{Q}^{1}(F_{Q2}^{1})=1\cdot 2+5\cdot 1=7=$$
$$5F_{\times 8}^{4}[4F_{6}^{3}(3F_{4}^{2},2F_{3}^{2}),2F_{4}^{3}(4F_{3}^{2})]+2F_{\times 5}^{2}[5F_{4}^{3}(4F_{3}^{2})],$$

$$f_{\times}^{5}=f_{P}^{4}(F_{P5}^{3})f_{Q}^{1}(F_{Q2}^{1})=1\cdot 1=1=P_{10}^{5}.$$

According to the last equality, we obtain the structural formula of the product

$$P_{5}^{3}[5F_{4}^{3}(4F_{3}^{2})]\times Q_{2}^{1}=P_{10}^{4}\{5F_{8}^{4}[4F_{6}^{3}(3F_{4}^{2},2F_{3}^{2}),2F_{4}^{3}(4F_{3}^{2})],2F_{5}^{4}[5F_{4}^{3}(4F_{3}^{2})]\}. \tag{13}$$

Thus, the product of a 4*D*-simplex by a segment is a polytope 5*D* including 5 tetrahedral prisms (8) and 2 4-simplexes. We can call the polytope (13) a 5-simplex-prism. The image of 5 -simplex-prism is shown in Figure 2.

The Product of a 4*D*-Cross-Polytope by a Segment

The structural formula of a 4*D*-cross-polytope is $P_{8}^{3}[16F_{4}^{3}(4F_{3}^{2})]$. It is necessary to find the structural formula of the product $P_{8}^{3}[16F_{4}^{2}(4F_{3}^{2})]\times Q_{2}^{1}$. According to theorem 1 and formulas (5), we find

Figure 2. 5-simplex-prism

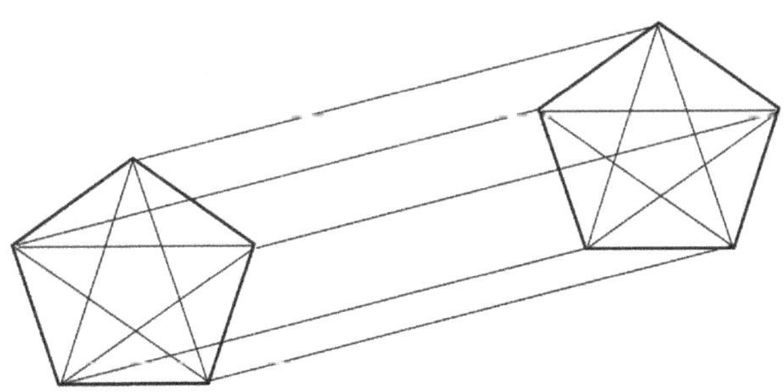

$$f_{\times}^{0} = f_{P}^{0} f_{Q}^{0} = 8 \cdot 2 = 16,$$

$$f_{\times}^{1} = f_{P}^{1} f_{Q}^{0} + f_{P}^{0} f_{Q}^{1} = 24 \cdot 2 + 8 \cdot 1 = 56,.$$

$$f_{\times}^{2} = f_{P}^{2}(F_{P3}^{2}) f_{Q}^{0}(F_{Q1}^{0}) + f_{P}^{1}(F_{P2}^{1}) f_{Q}^{1}(F_{Q2}^{1}) = 32 \cdot 2 + 24 \cdot 1 = 88 = 64F_{\times 6}^{2} + 24F_{\times 4}^{2},$$

$$f_{\times}^{3} = f_{P}^{3}(F_{P4}^{2}) f_{Q}^{0}(F_{Q1}^{0}) + f_{P}^{2}(F_{P3}^{2}) f_{Q}^{1}(F_{Q2}^{1}) = 16 \cdot 2 + 32 \cdot 1 =$$
$$64 = 32F_{\times 4}^{3}(4F_{3}^{2}) + 32F_{\times 6}^{2}(3F_{4}^{2}, 2F_{3}^{2}),$$

$$f_{\times}^{4} = f_{P}^{4}\{\mathrm{F}_{8}^{4}[16F_{P4}^{3}(4F_{3}^{3})]\} f_{Q}^{0}(F_{Q1}^{0}) + f_{P}^{2}(F_{P4}^{2}) f_{Q}^{1}(F_{Q2}^{1}) = 1 \cdot 2 + 16 \cdot 1 = 18 =$$
$$16F_{\times 8}^{4}[4F_{6}^{3}(3F_{4}^{2}, 2F_{3}^{2}), 2F_{4}^{3}(4F_{3}^{2})] + 2F_{\times 8}^{2}[16F_{4}^{3}(4F_{3}^{2})],$$

$$f_{\times}^{5} = f_{P}^{4}\{\mathrm{F}_{4}^{3}[16F_{4}^{3}(4F_{3}^{3}) f_{Q}^{1}(F_{Q2}^{1}) = 1 \cdot 1 = 1 = P_{16}^{5}.$$

According to the last equality, we obtain the structural formula of the product

$$P_{5}^{3}[5F_{8}^{3}[16F_{4}^{3}(4F_{3}^{2})] \times Q_{2}^{1} =$$
$$P_{16}^{4}\{16F_{8}^{4}[4F_{6}^{3}(3F_{4}^{2}, 2F_{3}^{2}), 2F_{4}^{3}(4F_{3}^{2})], 2F_{8}^{4}[16F_{4}^{3}(4F_{3}^{2})]\}. \quad (14)$$

Thus, the product of a $4D$-cross-polytope by a segment is a polytope $5D$ including 16 tetrahedral prisms (8) and 2 4-cross-polytopes. We can call the polytope (14) a 5-cross-prism.

The General Structural Formula of Polytopes Product by a Segment

We'll call prisms with bases in the form of polytopes the polytopic prisms.

Theorem 2

The product of a polytope by a segment, if all its faces of dimension 2 are equal, is a polytopic prism with the structure defined by equation

$$P^n_{K_n}\{f_{n-1}F^{n-1}_{K_{n-1}}...[f_3 F^3_{K_3}(f_2F^2_{K_2})]...\}\times Q^1_2 =$$
$$P^{n+1}_{2K_n}\{f_{n-1}F^n_{2K_{n-1}}...[f_3F^4_{2K_3}(f_2F^3_{2K_2},2F^2_{K_2})2F^3_{K_3}]...2F^n_{K_n}\}. \quad (15)$$

There $P^n_{K_n}$ is the polytope of dimension n with the number of vertices K_n; $F^{n-1}_{K_{n-1}}$ are hyper-faces with the number f_{n-1}, each having K_{n-1} vertices; accordingly $F^2_{K_2}$ are hyper-faces of dimension 2 at faces $F^3_{K_3}$; f_2 is the number of faces $F^2_{K_2}$ with the number of vertices K_2 and so on; $P^{n+1}_{2K_n}$ is a polytopic prism of dimension $n+1$ with the number of vertices $2K_n$, the bases of which are the polytopes $P^n_{K_n}$. The polytopic prism $P^{n+1}_{2K_n}$ includes f_{n-1} polytopic prisms of dimension n and the number of the vertices $2K_{n-1}$, and also two polytopes $P^n_{K_n}$ which are the bases of polytopic prisms and so on, up to prisms $F^3_{2K_2}$ with a number f_2 and with bases $F^2_{K_2}$.

Proof

Theorem 1 follows from the carried out examinations of the product of polytopes of different dimensions by a segment. General laws which allowed us to formulate the theorem can be traced in all these products. Whatever the dimension n of the polytope was, at multiplying it by a segment of each vertex of the polytope, the segment which is multiplied by a polytope as if grows. This forms polytopic prism, the bases of which are two copies of the initial polytope. The dimension of this polytopic prism is by 1 more than the dimension of the initial polytope, i.e. $n + 1$. This polytopic prism includes polytopic prisms built on the hyper-faces of the initial polytope and their number is equal to the number of hyper-faces of the initial polytope f_{n-1}. The dimension of these politopic prisms is n, since the dimension of the initial polytope hyper-faces is equal to $n - 1$, but when multiplied by a segment the dimension is increased by 1. In addition, a polytopic prism of dimension $n + 1$ consists of two copies of the initial polytope of dimension n, as the bases of the prism. In its turn, the polytopic prisms of dimension n include the polytopic prisms of dimension $n - 1$, built on the hyper-faces of the polytope of dimension $n - 1$, and two copies of the polytope of dimension $n - 1$. This process of the polytopic prisms inclusion in each other happens until the formation of three-dimensional prisms built on two-dimensional facets of the initial polytope.

The equality (15) is an analytical expression of these constructions. This process of formation occurs when all two-dimensional faces of the initial polytope are the same, i.e. they are equal polygons. If the initial polytope has two-dimensional faces with different numbers of vertices so the analytical record of these polytopes product by a segment will be added by prisms built on different bases. The ratios (15) don't include such options of polytopic prisms.

Theorem 2 it is proved.

Supplement. Values of f_i in (15) depend on the form and dimension of the polytope. According to Zhizhin (2014) the formulas for their calculation in the case of a simplex, cube and cross-polytope will be the following: for simplex $f_i = C_{i+1}^{i} = i+1$; for cube $f_i = 2C_{i}^{i-1} = 2i$; for cross-polytope $f_i = 2^i, i > 3, f_3 = 4$.

Naturally, the results of research about products of polytopes by a segment are to be taken into account when analyzing the products of polytopes by more complex figures, since they all have edges are factors in these products.

THE PRODUCTS OF POLYTOPES BY A TRIANGLE

The Product of a Triangle by a Triangle

There we consider the product $P_3^2 \times Q_3^2$. According to theorem 1 and ratios (5) we have

$$f_\times^0 = f_P^0 f_Q^0 = 3 \cdot 3 = 9,$$

$$f_\times^1 = f_P^1 f_Q^0 + f_P^0 f_Q^1 = 3 \cdot 3 + 3 \cdot 3 = 18,.$$

$$f_\times^2 = f_P^2(4F_{P3}^2) f_Q^0(F_{Q1}^0) + f_P^1(F_{P2}^1) f_Q^1(F_{Q2}^1) + f_P^0(F_{P1}^0) f_Q^2(F_{Q3}^2) = \\ 1 \cdot 3 + 3 \cdot 3 + 3 \cdot 1 = 15 = 6F_{\times 3}^2 + 9F_{\times 4}^2,$$

$$f_\times^3 = f_P^3(F_{P3}^2) f_Q^1(F_{Q2}^1) + f_P^1(F_{P2}^1) f_Q^2(F_{Q3}^2) = 1 \cdot 3 + 3 \cdot 1 = 6 = 6F_{\times 6}^3,$$

$$f_\times^4 = f_P^2(F_{P3}^2) f_Q^2(F_{Q3}^2) = 1 \cdot 1 = 1 = P_{\times 9}^4.$$

Thus, the product of a triangle by a triangle is $4D$-polytope having 9 vertices and 6 triangle prisms. The existence of such polytope was proved of Ziegler (1995) independently.

In particular, the fundamental difference between this polytope and simplicial polytopes and between this polytope and three-dimensional polyhedrons because of its multidimensionality is indicated of Ziegler (1995). The structural formula of this polytope (the triangular prismahedron) has the form

$$P_3^2(F_{P3}^2)\times Q_3^2(F_{Q3}^2) = P_9^4[6P_6^3(3F_4^2, F_3^2)]. \tag{16}$$

The image of a triangular prismahedron (Zhizhin, 2017) is shown in Figure 3.

The Product of a Polygon by a Triangle

According to theorem 1 and ratios (5) for the product $P_i^2(F_{Pi}^2)\times Q_3^2(F_{Q3}^2)$ (i is the number of the polygon sides) we have

$$f_\times^0 = f_P^0 f_Q^0 = i \cdot 3 = 3i,$$

$$f_\times^1 = f_P^1 f_Q^0 + f_P^0 f_Q^1 = i\cdot 3 + 3\cdot i = 6i,.$$

Figure 3. The triangular prismahedron

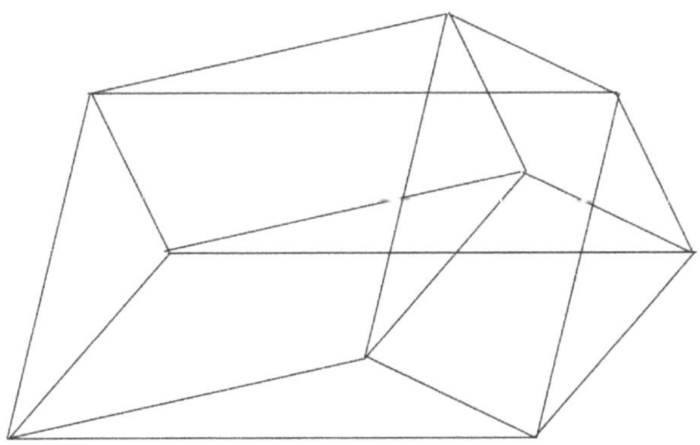

$$f_{\times}^{2} = f_{P}^{2}(F_{Pi}^{2})f_{Q}^{0}(F_{Q1}^{0}) + f_{P}^{1}(F_{P2}^{1})f_{Q}^{1}(F_{Q2}^{1}) + f_{P}^{0}(F_{P1}^{0})f_{Q}^{2}(F_{Q3}^{2}) =$$
$$1 \cdot 3 + i \cdot 3 + i \cdot 1 = 3 + 4i = 3F_{\times i}^{2} + 3iF_{\times 4}^{2} + iF_{\times 3}^{2},$$

$$f_{\times}^{3} = f_{P}^{3}(F_{Pi}^{2})f_{Q}^{1}(F_{Q2}^{1}) + f_{P}^{1}(F_{P2}^{1})f_{Q}^{2}(F_{Q3}^{2}) = 1 \cdot 3 + i \cdot 1 = 3 + i = iF_{\times 6}^{3} + 3F_{\times 2i}^{3},$$

$$f_{\times}^{4} = f_{P}^{2}(F_{Pi}^{2})f_{Q}^{2}(F_{Q3}^{2}) = 1 \cdot 1 = 1 = P_{\times 3i}^{4}.$$

Thus, the product of the polygon F_{i}^{2} by the triangle F_{3}^{2} is a 4-polytope having $3i$ vertices, 3 prisms with bases F_{i}^{2} and i prisms with triangle bases. The structural formula of this product has the form

$$P_{i}^{3}(F_{Pi}^{2}) \times Q_{3}^{2}(F_{Q3}^{2}) = P_{3i}^{4}[3F_{2i}^{3}(\mathrm{i}\,F_{4}^{2}, 2F_{3}^{2}), \mathrm{i}\,F_{6}^{3}(3F_{4}^{2}, 2F_{3}^{2})]. \quad (17)$$

Examples of products of a square, a pentagon and a hexagon by a triangle are shown in Figures 4 – 6.

*Figure 4. The 4*3 – angular prismahedron*

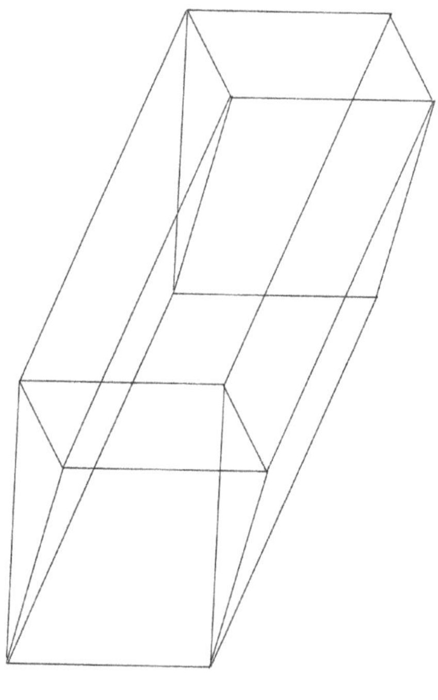

*Figure 5. The 5*3 – angular prismahedron*

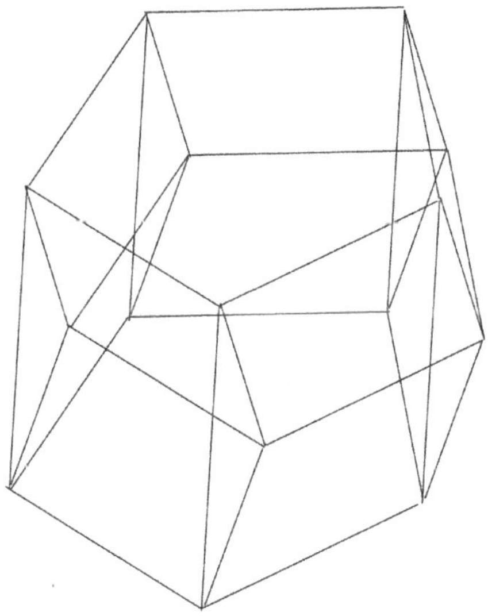

*Figure 6. The 6*3 – angular prismahedron*

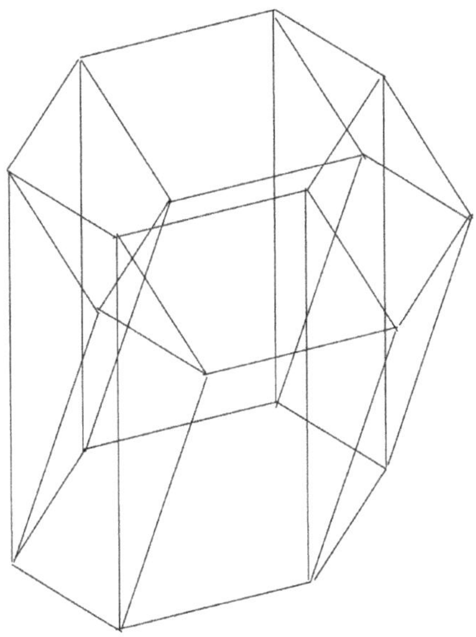

The polytope (17) is a generalization of the polytope (16) in the case when one of the multiplied polygons has the number of sides greater than 3. This polytope in the general case can be called 3 *i - angular prismahedron. Therefore, the polytope (16) can be more accurately described as 3*3 - angular prismahedron.

The Product of a Tetrahedron by a Triangle

According to theorem 1 and ratios (5) for the numbers of elements of different dimension in the product $P_4^2(4F_3^2)\times Q_3^2$ we have

$$f_\times^0 = f_P^0 f_Q^0 = 4\cdot 3 = 12,$$

$$f_\times^1 = f_P^1 f_Q^0 + f_P^0 f_Q^1 = 6\cdot 3 + 3\cdot 4 = 30,.$$

$$f_\times^2 = f_P^2(F_{P3}^2) f_Q^0(F_{Q1}^0) + f_P^1(F_{P2}^1) f_Q^1(F_{Q2}^1) + f_P^0(F_{P1}^0) f_Q^2(F_{Q3}^2) =$$
$$4\cdot 1 + 6\cdot 3 + 4\cdot 1 = 26 = 8F_{\times 3}^2 + 18F_{\times 4}^2,$$

$$f_\times^3 = f_P^3(F_{P4}^2) f_Q^0(F_{Q1}^0) + f_P^2(F_{P3}^2) f_Q^1(F_{Q2}^1) + f_P^1(F_{P2}^1) f_Q^2(F_{Q3}^2) =$$
$$1\cdot 3 + 4\cdot 1 + 6\cdot 1 = 21 = 18F_{\times 6}^3 + 3F_{\times 4}^3,$$

$$f_\times^4 = f_P^3(F_{P4}^3) f_Q^1(F_{Q2}^1) + f_P^2(F_{P3}^2) f_Q^2(F_{Q3}^2) = 1\cdot 3 + 4\cdot 1 = 7 = 3F_{\times 8}^4 + 4F_{\times 9}^4,$$
$$f_\times^5 = f_P^3(F_{P4}^3) f_Q^2(F_{Q3}^2) = 1\cdot 1 = 1 = P_{\times 12}^5.$$

Thus, the product of a tetrahedron by a triangle is a 5-polytope consisting from 3 tetrahedron prisms (8) and 4 triangular prismahedrons (16). The structural formula of this polytope has the form

$$P_4^3(4F_3^2)]\times Q_3^2 = P_{12}^5\{4F_9^4[6F_6^3(3F_4^2,2F_3^2),2F_8^4[4F_6^3(3F_4^2,2F_3^2),2F_4^3(4F_3^2)]\}. \qquad (18)$$

We can call the polytope (18) a tetrahedral prismahedron. The image of this polytope on a two-dimension plane is shown in Figure 7.

Figure 7. The tetrahedral prismahedron

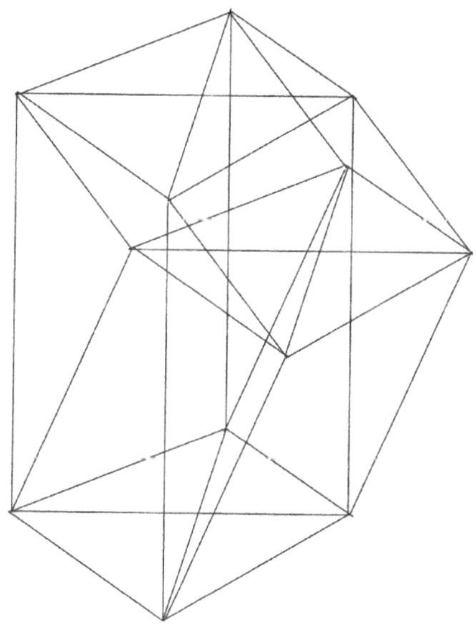

The Product of an Octahedron by a Triangle

According to theorem 1 and ratios (5) for the numbers of elements of different dimension in the product $P_6^2(8F_3^2)\times Q_3^2$ we have

$$f_\times^0 = f_P^0 f_Q^0 = 6 \cdot 3 = 18,$$

$$f_\times^1 = f_P^1 f_Q^0 + f_P^0 f_Q^1 = 12 \cdot 3 + 6 \cdot 3 = 54,.$$

$$f_\times^2 = f_P^2(F_{P3}^2) f_Q^0(F_{Q1}^0) + f_P^1(F_{P2}^1) f_Q^1(F_{Q2}^1) + f_P^0(F_{P1}^0) f_Q^2(F_{Q3}^2) = 8 \cdot 1 + 12 \cdot 3 + 6 \cdot 1 = 66 = 30F_{\times3}^2 + 36F_{\times4}^2,$$

$$f_\times^3 = f_P^3(F_{P6}^2) f_Q^0(F_{Q1}^0) + f_P^2(F_{P3}^2) f_Q^1(F_{Q2}^1) + f_P^1(F_{P2}^1) f_Q^2(F_{Q3}^2) = 1 \cdot 3 + 8 \cdot 3 + 12 \cdot 1 = 39 = 36F_{\times6}^3(3F_4^2, 2F_3^2) + 3F_{\times6}^3(8F_3^2),$$

$$
\begin{aligned}
f_{\times}^{4} &= f_{P}^{3}[F_{P6}^{3}(8F_{3}^{2})]f_{Q}^{1}(F_{Q2}^{1}) + f_{P}^{2}(F_{P3}^{2})f_{Q}^{2}(F_{Q3}^{2}) = 1\cdot 3 + 8\cdot 1 = 11 \\
&= 3F_{\times 12}^{4}[8F_{6}^{3}(3F_{4}^{2},2F_{3}^{2}),2F_{6}^{3}(8F_{3}^{2})] + 8F_{\times 9}^{4}[6F_{6}^{3}(3F_{4}^{2},2F_{3}^{2})], \\
f_{\times}^{5} &= f_{P}^{3}[F_{P6}^{3}(8F_{3}^{2})]f_{Q}^{2}(F_{Q3}^{2}) = 1\cdot 1 = 1 = P_{\times 18}^{5}.
\end{aligned}
$$

Thus, the product of an octahedron by a triangle is a 5-polytope including 8 3*3-angular prismahedrons (16) and 3 octahedral prisms (10).

The structural formula of the polytope P_{18}^{5} has the form

$$
P_{6}^{3}(8F_{3}^{2})]\times Q_{3}^{2} = P_{18}^{5}\{8F_{9}^{4}[6F_{6}^{3}(3F_{4}^{2},2F_{3}^{2}),3F_{12}^{3}[8F_{6}^{3}(3F_{4}^{2},2F_{3}^{2}),2F_{6}^{3}(8F_{3}^{2})]\}. \tag{19}
$$

We can call the polytope (19) an octahedral prismahedron. The image of the polytope P_{18}^{5} is shown in Figure 8.

The Product of a Cube by a Triangle

According to theorem 1 and ratios (5) for the numbers of elements of different dimension in the product $P_{8}^{2}(6F_{4}^{2})\times Q_{3}^{2}$ we have

$$
f_{\times}^{0} = f_{P}^{0}f_{Q}^{0} = 8\cdot 3 = 24,
$$

$$
f_{\times}^{1} = f_{P}^{1}f_{Q}^{0} + f_{P}^{0}f_{Q}^{1} = 12\cdot 3 + 8\cdot 3 = 60,.
$$

Figure 8. The octahedral prismahedron

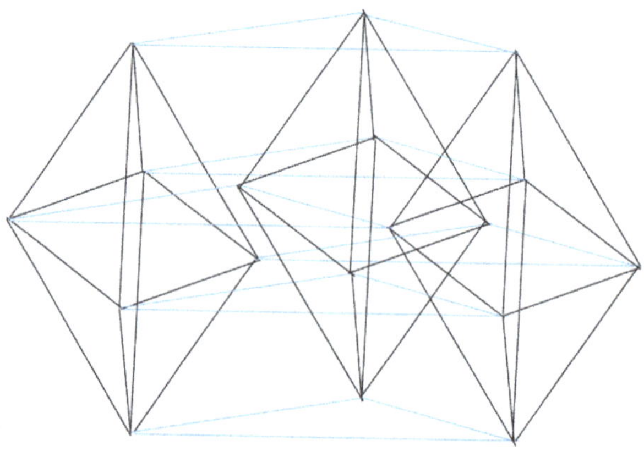

$$f_\times^2 = f_P^2(F_{P4}^2)f_Q^0(F_{Q1}^0) + f_P^1(F_{P2}^1)f_Q^1(F_{Q2}^1) + f_P^0(F_{P1}^0)f_Q^2(F_{Q3}^2) =$$
$$6 \cdot 3 + 12 \cdot 3 + 8 \cdot 1 = 62 = 8F_{\times 3}^2 + 54F_{\times 4}^2,$$

$$f_\times^3 = f_P^3(F_{P8}^2)f_Q^0(F_{Q1}^0) + f_P^2(F_{P4}^2)f_Q^1(F_{Q2}^1) + f_P^1(F_{P2}^1)f_Q^2(F_{Q3}^2) =$$
$$1 \cdot 3 + 6 \cdot 3 + 12 \cdot 1 = 33 = 21F_{\times 8}^3 + 12F_{\times 6}^3,$$

$$f_\times^4 = f_P^3(F_{P8}^3)f_Q^1(F_{Q2}^1) + f_P^2(F_{P3}^2)f_Q^2(F_{Q3}^2) = 1 \cdot 3 + 6 \cdot 1 = 9$$
$$= 3F_{16}^4 + 6F_{12}^4,$$

$$f_\times^5 = f_P^3(F_{P8}^3)f_Q^2(F_{Q3}^2) = 1 \cdot 1 = 1 = P_{\times 24}^5.$$

Thus, the product of a cube by a triangle is a 5 - polytope including 3 hyper-cubes (9) and 6 3*4-angular prismahedrons (17).

The structural formula of the polytope P_{24}^5 has the form

$$P_{24}^5\{3F_{16}^4(80F_4^2), 6F_{12}^4[3F_8^3(4F_4^2, 2F_3^2), 4F_6^3(3F_4^2, 2F_3^2)]\}. \tag{20}$$

We can call the polytope (20) a cube-polytopic prismahedron. The image of the polytope P_{24}^5 is shown in Figure 9.

The Product of a Dodecahedral by a Triangle

According to theorem 1 and ratios (5) for the numbers of elements of different dimension in the product $P_{20}^3(12F_5^2) \times Q_3^2$ we have

Figure 9. The cube-polytopic prismahedron

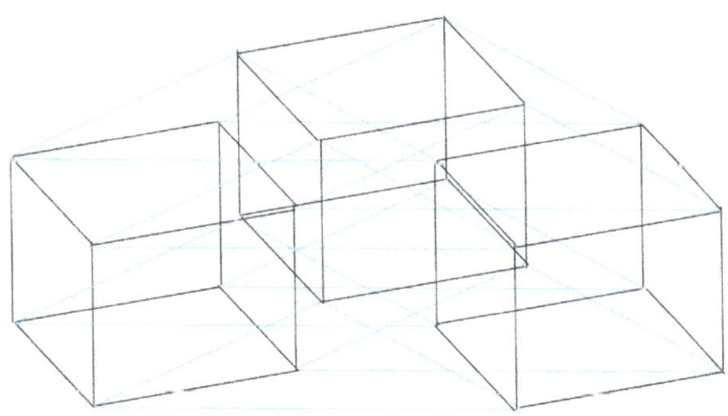

$$f_{\times}^{0} = f_P^0 f_Q^0 = 20 \cdot 3 = 60,$$

$$f_{\times}^{1} = f_P^1 f_Q^0 + f_P^0 f_Q^1 = 30 \cdot 3 + 20 \cdot 3 = 150,.$$

$$f_{\times}^{2} = f_P^2(F_{P5}^2) f_Q^0(F_{Q1}^0) + f_P^1(F_{P2}^1) f_Q^1(F_{Q2}^1) + f_P^0(F_{P1}^0) f_Q^2(F_{Q3}^2)$$
$$= 12 \cdot 3 + 30 \cdot 3 + 20 \cdot 1 = 146 = 36F_{\times 5}^2 + 90F_{\times 4}^2 + 20F_{\times 3}^2,$$

$$f_{\times}^{3} = f_P^3(F_{P20}^3) f_Q^0(F_{Q1}^0) + f_P^2(F_{P5}^2) f_Q^1(F_{Q2}^1) + f_P^1(F_{P2}^1) f_Q^2(F_{Q3}^2)$$
$$= 1 \cdot 3 + 12 \cdot 3 + 30 \cdot 1 = 69 = 3F_{\times 20}^3 + 36F_{\times 10}^3 + 30F_{\times 6}^3,$$

$$f_{\times}^{4} = f_P^3(F_{P20}^3) f_Q^1(F_{Q2}^1) + f_P^2(F_{P5}^2) f_Q^2(F_{Q3}^2) = 1 \cdot 3 + 12 \cdot 1 = 15$$
$$= 3F_{\times 40}^4 + 12F_{\times 15}^4,$$

$$f_{\times}^{5} = f_P^3(F_{P20}^3) f_Q^2(F_{Q3}^2) = 1 \cdot 1 = 1 = P_{\times 60}^5.$$

Thus, the product of a dodecahedron by a triangle is a 5-polytope including 3 dodecahedral prisms (12) and 12 3*5-angular prismahedrons (17).

The structural formula of the polytope P_{60}^5 has the form

$$P_{60}^5\{3F_{40}^4[12F_{10}^3(5F_4^2, 2F_3^2), 2F_{20}^3(12F_5^2)], 12F_{15}^4[3F_{10}^3(5F_4^2, 2F_5^2), 5F_6^3(3F_4^2, 2F_3^2)]\}. \qquad (21)$$

We can call the polytope (21) a dodeca - polytopic prismahedron.

The Product of an Icosahedron by a Triangle

According to theorem 1 and ratios (5) for the numbers of elements of different dimension in the product $P_{12}^3(20F_5^2) \times Q_3^2$ we have

$$f_{\times}^{0} = f_P^0 f_Q^0 = 12 \cdot 3 = 36,$$

$$f_{\times}^{1} = f_P^1 f_Q^0 + f_P^0 f_Q^1 = 30 \cdot 3 + 12 \cdot 3 = 126,.$$

$$f_\times^2 = f_P^2(F_{P3}^2)f_Q^0(F_{Q1}^0) + f_P^1(F_{P2}^1)f_Q^1(F_{Q2}^1) + f_P^0(F_{P1}^0)f_Q^2(F_{Q3}^2)$$
$$= 20 \cdot 3 + 30 \cdot 3 + 12 \cdot 1 = 162 = 90F_{\times 4}^2 + 72F_{\times 3}^2,$$

$$f_\times^3 = f_P^3(F_{P12}^3)f_Q^0(F_{Q1}^0) + f_P^2(F_{P3}^2)f_Q^1(F_{Q2}^1) + f_P^1(F_{P2}^1)f_Q^2(F_{Q3}^2)$$
$$= 1 \cdot 3 + 20 \cdot 3 + 30 \cdot 1 = 93 = 3F_{\times 12}^3 + 90F_{\times 6}^3,$$

$$f_\times^4 = f_P^3(F_{P12}^3)f_Q^1(F_{Q2}^1) + f_P^2(F_{P3}^2)f_Q^2(F_{Q3}^2) = 1 \cdot 3 + 20 \cdot 1 = 23$$
$$= 3F_{\times 24}^4 + 20F_{\times 9}^4,$$

$$f_\times^5 - f_P^3(F_{P12}^3)f_Q^2(F_{Q3}^2) = 1 \cdot 1 = 1 = P_{\times 36}^5.$$

Thus, the product of an icosahedron by a triangle is a 5-polytope including 3 icosahedral prisms (11) and 20 3*3-angular prismahedrons (17).

The structural formula of the polytope P_{36}^5 has the form

$$P_{36}^5\{20F_9^4[6F_6^3(3F_4^2,2F_3^2)],3F_{24}^4[20F_6^3(3F_4^2,2F_3^2),2F_{12}^3(20F_3^2)]\}. \tag{22}$$

We can call the polytope (22) an icosa - polytopic prismahedron.

The Product of a 4-Simplex by a Triangle

According to theorem 1 and ratios (5) for the numbers of elements of different dimension in the product $P_5^3[5F_4^3(4F_3^2)] \times Q_3^2$ we have

$$f_\times^0 = f_P^0 f_Q^0 = 5 \cdot 3 = 15,$$

$$f_\times^1 = f_P^1 f_Q^0 + f_P^0 f_Q^1 = 10 \cdot 3 + 5 \cdot 3 = 45,$$

$$f_\times^2 = f_P^2(F_{P3}^2)f_Q^0(F_{Q1}^0) + f_P^1(F_{P2}^1)f_Q^1(F_{Q2}^1) + f_P^0(F_{P1}^0)f_Q^2(F_{Q3}^2)$$
$$= 10 \cdot 3 + 10 \cdot 3 + 5 \cdot 1 = 65 = 30F_{\times 4}^2 + 35F_{\times 3}^2,$$

$$f_\times^3 = f_P^3(F_{P4}^3)f_Q^0(F_{Q1}^0) + f_P^2(F_{P3}^2)f_Q^1(F_{Q2}^1) + f_P^1(F_{P2}^1)f_Q^2(F_{Q3}^2)$$
$$= 5 \cdot 3 + 10 \cdot 3 + 10 \cdot 1 = 55 = 15F_{\times 4}^3 + 40F_{\times 6}^3,$$

$$f_{\times}^{4} = f_{P}^{4}(F_{P5}^{4})f_{Q}^{0}(F_{Q1}^{0}) + f_{P}^{3}(F_{P4}^{3})f_{Q}^{1}(F_{Q2}^{1}) + f_{P3}^{2}(F_{P}^{2})f_{Q}^{2}(F_{Q3}^{2})$$
$$= 1\cdot 3 + 5\cdot 3 + 10\cdot 1 = 28 = 3F_{\times 5}^{4} + 15F_{\times 8}^{4} + 10F_{\times 9}^{4},$$

$$f_{\times}^{5} = f_{P}^{4}(F_{P5}^{4})f_{Q}^{1}(F_{Q2}^{1}) + f_{P}^{3}(F_{P4}^{3})f_{Q}^{2}(F_{Q3}^{2}) = 1\cdot 3 + 5\cdot 1 = 8$$
$$= 3F_{\times 10}^{5} + 5F_{\times 12}^{5},$$

$$f_{\times}^{6} = f_{P}^{4}(F_{P5}^{4})f_{Q}^{2}(F_{Q3}^{2}) = 1\cdot 1 = 1 = P_{\times 15}^{6}.$$

Thus, the product of a 4-simplex by a triangle is a 6-polytope including 3 5-simplex-polytopic prisms (13) and 5 tetrahedral prismahedrons (18).

The structural formula of the polytope P_{15}^{6} has the form

$$P_{15}^{6}\{3F_{10}^{5}[5F_{8}^{4}[4F_{4}^{3}(3F_{4}^{2},2F_{3}^{2}),2F_{4}^{3}(4F_{3}^{2})],2F_{5}^{4}[5F_{4}^{3}(4F_{3}^{2})]],$$
$$5F_{12}^{5}[4F_{9}^{4}[6F_{6}^{3}(3F_{4}^{2},2F_{3}^{2}),3F_{8}^{4}[4F_{6}^{3}(3F_{4}^{2},2F_{3}^{2}),2F_{4}^{3}(4F_{3}^{2})]]\}. \qquad (23)$$

We can call the polytope (23) a 6-simplex-polytopic prismahedron. The image of this polytope is shown in Figure 10.

The Product of a 4-Cross-Polytope by a Triangle

According to theorem 1 and ratios (5) for the numbers of elements of different dimension in the product $P_{8}^{3}[16F_{4}^{3}(4F_{3}^{2})]\times Q_{3}^{2}$ we have

Figure 10. The 6-sinplex-polytopic prismahedron

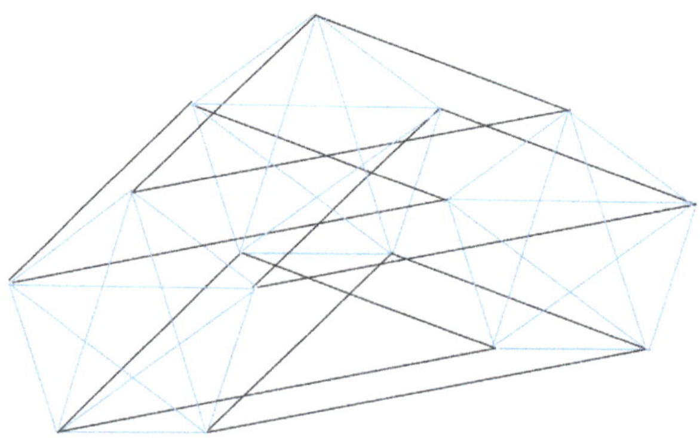

$$f_{\times}^{0} = f_{P}^{0} f_{Q}^{0} = 8 \cdot 3 = 24,$$

$$f_{\times}^{1} = f_{P}^{1} f_{Q}^{0} + f_{P}^{0} f_{Q}^{1} = 24 \cdot 3 + 8 \cdot 3 = 96,$$

$$f_{\times}^{2} = f_{P}^{2}(F_{P3}^{2}) f_{Q}^{0}(F_{Q1}^{0}) + f_{P}^{1}(F_{P2}^{1}) f_{Q}^{1}(F_{Q2}^{1}) + f_{P}^{0}(F_{P1}^{0}) f_{Q}^{2}(F_{Q3}^{2})$$
$$= 32 \cdot 3 + 24 \cdot 3 + 8 \cdot 1 = 176 = 72F_{\times 4}^{2} + 104F_{\times 3}^{2},$$

$$f_{\times}^{3} = f_{P}^{3}(F_{P4}^{3}) f_{Q}^{0}(F_{Q1}^{0}) + f_{P}^{2}(F_{P3}^{2}) f_{Q}^{1}(F_{Q2}^{1}) + f_{P}^{1}(F_{P2}^{1}) f_{Q}^{2}(F_{Q3}^{2})$$
$$= 16 \cdot 3 + 32 \cdot 3 + 24 \cdot 1 = 168 = 48F_{\times 4}^{3} + 120F_{\times 6}^{3},$$

$$f_{\times}^{4} = f_{P}^{4}(F_{P8}^{4}) f_{Q}^{0}(F_{Q1}^{0}) + f_{P}^{3}(F_{P4}^{3}) f_{Q}^{1}(F_{Q2}^{1}) + f_{P3}^{2}(F_{P}^{2}) f_{Q}^{2}(F_{Q3}^{2})$$
$$= 1 \cdot 3 + 16 \cdot 3 + 32 \cdot 1 = 83 = 51F_{\times 8}^{4} + 32F_{\times 9}^{4},$$

$$f_{\times}^{5} = f_{P}^{4}(F_{P8}^{4}) f_{Q}^{1}(F_{Q2}^{1}) + f_{P}^{3}(F_{P4}^{3}) f_{Q}^{2}(F_{Q3}^{2}) = 1 \cdot 3 + 16 \cdot 1 = 19$$
$$= 3F_{\times 16}^{5} + 16F_{\times 12}^{5},$$

$$f_{\times}^{6} = f_{P}^{4}(F_{P8}^{4}) f_{Q}^{2}(F_{Q3}^{2}) = 1 \cdot 1 = 1 = P_{\times 24}^{6}.$$

Thus, the product of a 4-cross-polytope by a triangle is a 6-polytope including 3 5-crosspolytopic prisms (10) and 16 tetrahedral prismahedrons (3).

The structural formula of the polytope P_{24}^{6} has the form

$$P_{24}^{6}\{3F_{16}^{5}[16F_{8}^{4}[4F_{6}^{3}(3F_{4}^{2},2F_{3}^{2}),2F_{4}^{3}(4F_{3}^{2})],2F_{8}^{4}[16F_{4}^{3}(4F_{3}^{2})]],$$
$$16F_{12}^{5}[4F_{9}^{4}[6F_{6}^{3}(3F_{4}^{2},2F_{3}^{2}),3F_{8}^{4}[4F_{6}^{3}(3F_{4}^{2},2F_{3}^{2}),2F_{4}^{3}(4F_{3}^{2})]]\}.$$

We can call the polytope a 6-cross - polytopic prismahedron.

The General Structural Formula for the Product of Polytopes by a Triangle

If we pay attention to a polytope taking into account its dimensions and the number of vertices and the number of its facets and the number of vertices of every facet, we can prove the general rule of changing these parameters of a polytope for its multiplication by a triangle.

Theorem 3

Let's assume that a convex polytope of dimension *n,* with the number of vertices *m*, has a number of facets ν with vertex number *k* for every facet. Then its product by a triangle is defined by the equation

$$P_m^n(\nu F_k^{n-1}) \times Q_3^2 = P_{3m}^{n+2}(\nu F_{3k}^{n+1}, 3F_{2m}^{n+1}). \tag{24}$$

Proof

When multiplying the polytope P_m^n by the triangle Q_3^2 the number of vertices of the polytope is multiplied by 3, since the triangle has 3 edges and each edge at this multiplication translates polytope vertices in its direction. At that the dimension of the polytope naturally, according to the definition (Ziegler, 1995), is increased by two. Thus, we'll get a polytope P_{3m}^{n+2}. If we multiply the polytope by the triangle, its facets F_k^{n-1} are multiplied by the triangle also. The number of these facets of the product is equal to the number of facets of the polytope P_m^n, i.e. it is equal to ν. In addition, at the multiplication by the triangle the polytope P_m^n is multiplied by each edge (i.e. multiplication by Q_2^1 3 times). Therefore, 3 facets F_{2m}^{n+1} more appear as facets of the polytope P_{3m}^{n+2}. It is easy to make sure that all the considered above products of the polytopes by the triangle satisfy the equality (24).

Theorem 3 is proved.

Corollary

If the polygon Q_t^2 with the number of vertices *t* is the polytope factor in (24), the equation (24) can be rewritten as

$$P_m^n(\nu F_k^{n-1}) \times Q_t^2 = P_{tm}^{n+2}(\nu F_{tk}^{n+1}, t\, F_{2m}^{n+1}). \tag{25}$$

As an example, the result of the product of a tetrahedron by a square is given in Figure 11.

THE PRODUCTS OF POLYTOPES BY A TETRAHEDRON

The Product of a Tetrahedron by a Tetrahedron

According to theorem 1 and ratios (5) for the numbers of elements of different dimension in the product $P_4^3(4F_3^2)\times Q_3^2(4F_3^2)$ we have

$$f_\times^0 = f_P^0 f_Q^0 = 4 \cdot 4 = 16,$$

$$f_\times^1 = f_P^1 f_Q^0 + f_P^0 f_Q^1 = 6 \cdot 4 + 4 \cdot 6 = 48,$$

$$f_\times^2 = f_P^2(F_{P3}^2) f_Q^0(F_{Q1}^0) + f_P^1(F_{P2}^1) f_Q^1(F_{Q2}^1) + f_P^0(F_{P1}^0) f_Q^2(F_{Q3}^2)$$
$$= 4 \cdot 4 + 6 \cdot 4 + 4 \cdot 4 = 68 = 36F_{\times4}^2 + 32F_{\times3}^2,$$

Figure 11. The product of a tetrahedron by a square

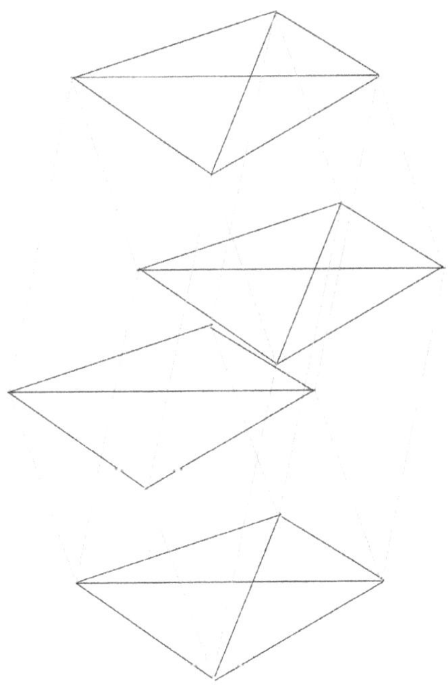

$$f_\times^3 = f_P^3(F_{P4}^3)f_Q^0(F_{Q1}^0) + f_P^2(F_{P3}^2)f_Q^1(F_{Q2}^1) + f_P^1(F_{P2}^1)f_Q^2(F_{Q3}^2) + f_P^0(F_{P1}^0)f_Q^3(F_{Q4}^3)$$
$$= 1\cdot 4 + 4\cdot 6 + 6\cdot 4 + 4\cdot 1 = 56 = 8F_{\times 4}^3 + 48F_{\times 6}^3,$$

$$f_\times^4 = f_P^3(F_{P4}^3)f_Q^1(F_{Q1}^1) + f_P^2(F_{P3}^2)f_Q^2(F_{Q3}^2) + f_P^1(F_{P2}^1)f_Q^3(F_{Q4}^3)$$
$$= 1\cdot 6 + 4\cdot 4 + 6\cdot 1 = 28 = 12F_{\times 8}^4 + 16F_{\times 9}^4,$$

$$f_\times^5 = f_P^3(F_{P4}^3)f_Q^2(F_{Q3}^2) + f_P^2(F_{P3}^2)f_Q^3(F_{Q4}^3) = 1\cdot 4 + 4\cdot 1 = 8 = 8F_{\times 12}^5,$$

$$f_\times^6 = f_P^3(F_{P4}^3)f_Q^2(F_{Q4}^3) = 1\cdot 1 = 1 = P_{\times 16}^6.$$

Thus, the product of a tetrahedron by a tetrahedron is a 6-polytope including 8 tetrahedral prismahedrons (18).

The short structural formula of the polytope P_{16}^6 has the form

$$P_{16}^6(8F_{12}^5). \tag{26}$$

The image of this polytope is shown in Figure 12.

Figure 12. The product of a tetrahedron by a tetrahedron

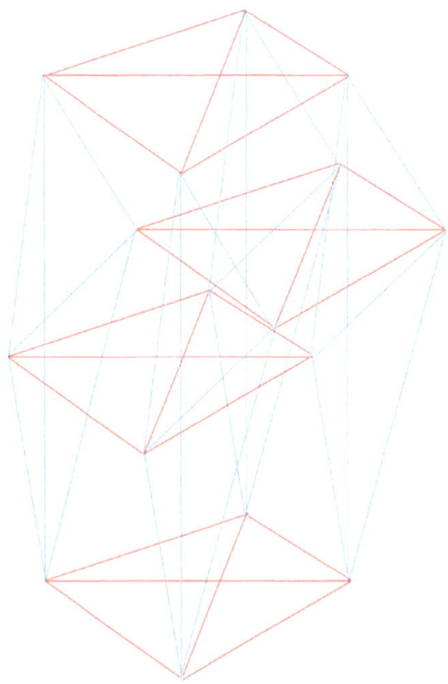

We can call the polytope (26) a 6-complex of tetrahedral prismahedrons.

The Product of an Octahedron by a Tetrahedron

According to theorem 1 and ratio (5) for the numbers of elements of different dimension in the product $P_6^3(8F_{P3}^2) \times Q_4^3(4F_{Q3}^2)$ we have

$$f_\times^0 = f_P^0 f_Q^0 = 6 \cdot 4 = 24,$$

$$f_\times^1 = f_P^1 f_Q^0 + f_P^0 f_Q^1 = 12 \cdot 4 + 6 \cdot 6 = 84,$$

$$f_\times^2 = f_P^2(F_{P3}^2) f_Q^0(F_{Q1}^0) + f_P^1(F_{P2}^1) f_Q^1(F_{Q2}^1) + f_P^0(F_{P1}^0) f_Q^2(F_{Q3}^2)$$
$$= 8 \cdot 4 + 12 \cdot 6 + 6 \cdot 4 = 128 = 72F_{\times 4}^2 + 56F_{\times 3}^2,$$

$$f_\times^3 = f_P^3(F_{P6}^3) f_Q^0(F_{Q1}^0) + f_P^2(F_{P3}^2) f_Q^1(F_{Q2}^1) + f_P^1(F_{P2}^1) f_Q^2(F_{Q3}^2) + f_P^0(F_{P1}^0) f_Q^3(F_{Q4}^3)$$
$$= 1 \cdot 4 + 8 \cdot 6 + 12 \cdot 4 + 6 \cdot 1 = 106 = 6F_{\times 4}^3 + 100F_{\times 6}^3,$$

$$f_\times^4 = f_P^3(F_{P6}^3) f_Q^1(F_{Q1}^1) + f_P^2(F_{P3}^2) f_Q^2(F_{Q3}^2) + f_P^1(F_{P2}^1) f_Q^3(F_{Q4}^3)$$
$$= 1 \cdot 6 + 8 \cdot 4 + 12 \cdot 1 = 50 = 12F_{\times 8}^4 + 32F_{\times 9}^4 + 6F_{\times 12}^4,$$

$$f_\times^5 = f_P^3(F_{P6}^3) f_Q^2(F_{Q3}^2) + f_P^2(F_{P3}^2) f_Q^3(F_{Q4}^3) = 1 \cdot 4 + 8 \cdot 1 = 12 = 8F_{\times 12}^5 + 4F_{\times 18}^5,$$

$$f_\times^6 = f_P^3(F_{P6}^3) f_Q^2(F_{Q4}^3) = 1 \cdot 1 = 1 = P_{\times 24}^6.$$

Thus, the product of an octahedron by a tetrahedron is a 6-polytope including 8 tetrahedral prismahedrons (18) and 4 octahedral polytopic prisms (19).

The short structural formula of the polytope P_{24}^6 has the form

$$P_{24}^6(4F_{18}^5, 8F_{12}^5). \tag{27}$$

The image of this polytope is shown in Figure 13.

We can call the polytope P_{24}^6 a 6-complex of tetrahedral prismahedrons and octahedral polytopic prisms.

Figure 13. The product of an octahedron by a tetrahedron

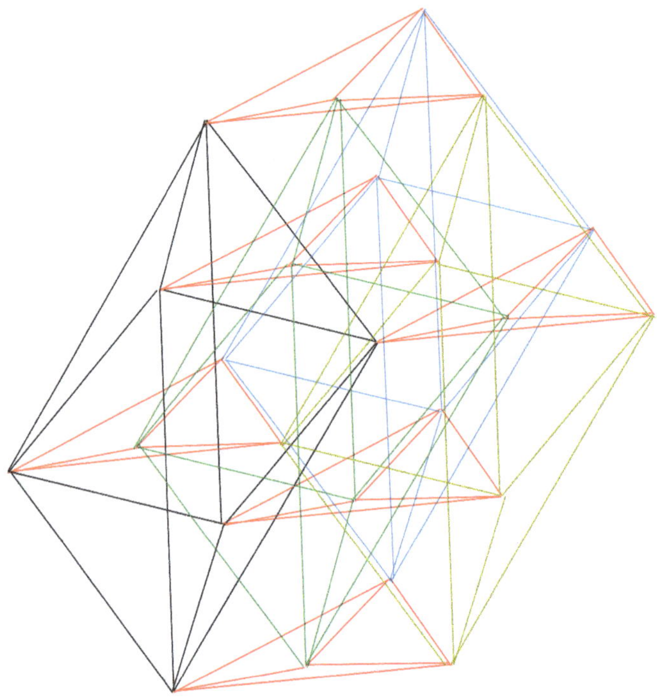

THE PRODUCT OF POLYTOPES WITH UNIFORM FACETS

Comparing (26) and (27) with equalities defined by theorems 1, 2 we note their general structure. This allows us to formulate and prove a general theorem on the product of polytopes with uniform facets.

Theorem 4

Let's assume that we have a convex polytope P of dimension n, with the number of vertices m, which has ν facets with the number of vertices k for each facet. Let's also set a convex polytope Q of dimension π, with the number of vertices l, having μ facets with the number of vertices λ for every facet. Then the product of these polytopes is defined by equation

$$P_m^n(\nu F_k^{n-1}) \times Q_l^\pi(\mu F_\lambda^{\pi-1}) = P_{lm}^{n+\pi}(\nu F_{lk}^{n+\pi-1}, \mu F_{\lambda m}^{n+\pi-1}). \tag{28}$$

Proof

For multiplication of the polytope P^n_m by the polytope Q^π_l the number of the vertices of polytope *P* is multiplied by the number of the vertices of polytope *Q* since each vertex of the polytope *P* at multiplication will correspond to all vertices of the polytope *Q*. The dimension of the product polytope according to the definition (Ziegler, 1995) is the sum of the dimensions of polytopes-factors, i.e. we receive the polytope $P^{n+\pi}_{lm}$ as the result of the product. Multiplication of the polytope *P* by the polytope *Q* is accompanied also by multiplication of the polytope *P* facets by the polytope *Q*. That is why the result of this multiplication will be the facets of the polytopes product defined according to the same rules, i.e. $F^{n+\pi-1}_{lk}$. The number of these facets is equal to the number of facets of the polytope *P*, i.e. ν. Besides, in the product of the polytopes *P* and *Q* there are facets formed by the product of the facets of polytope *Q* by polytope *P*. These are facets $F^{n+\pi-1}_{lk}$, their number is equal to the number of facets of the polytope *Q*, i.e. μ. The theorem 4 is proved.

Let's use theorem 4 for the product of an octahedron by a tetrahedron

$$P^3_6(8F^2_{P3}) \times Q^3_4(4F^2_{Q3}) = P^6_{24}(4F^5_{18}, 8F^5_{12}).$$

The received formula coincides with (27) got after detailed derivation.

It's easy to verify that all the previous formulas about polytopes product are particular cases of the formula (28). Theorem 4 allows to write down at once the result of the product of any convex polytopes with uniform faces without detailed derivation of this result. Let's build the multiplication table of existing polytopes of dimensions 3 and 4. For comfort of presentation the table is divided into a few tables (tables 1 – 4) to cover the products of all above-mentioned polytopes. The structure of new formed facets can be defined according to already mentioned formulas. The structural formulas in the table are given in a short form.

HIERARCHICAL AND TRANSLATIONAL FILLING SPACES

In Chapters 1 and 2 it showed that many molecules have a dimension greater than three if they are described in the form of geometric convex figures. Many of them have tetrahedral and octahedral coordination. But any substance is not

Table 1. Multiplication table of polytopes

Polytopes	Tetrahedron	Octahedron
Tetrahedron	$P_{16}^{6}(8F_{12}^{5})$	
Octahedron	$P_{24}^{6}(4F_{18}^{5},8F_{12}^{5})$	$P_{36}^{6}(16F_{18}^{5})$
Cube	$P_{32}^{6}(6F_{16}^{5},4F_{12}^{5})$	$P_{48}^{6}(6F_{24}^{5^{0}},8F_{24}^{5^{00}})$
Dodecahedron	$P_{80}^{6}(12F_{20}^{5},8F_{60}^{5})$	$P_{120}^{6}(12F_{130}^{5},8F_{60}^{5})$
Icosahedron	$P_{48}^{6}(20F_{12}^{5},4F_{36}^{5})$	$P_{72}^{6}(20F_{18}^{5},8F_{36}^{5})$
4-simplex	$P_{20}^{7}(5F_{16}^{5},4F_{15}^{5})$	$P_{30}^{7}(5F_{24}^{6^{0}},8F_{15}^{6})$
4-cube	$P_{64}^{7}(8F_{32}^{6},4F_{48}^{6})$	$P_{96}^{7}(8F_{48}^{6^{0}},8F_{48}^{6^{00}})$
4-cross-polytope	$P_{32}^{7}(16F_{16}^{6},4F_{24}^{6})$	$P_{48}^{7}(16F_{24}^{6^{0}},8F_{24}^{6^{00}})$

$F_{24}^{5^{0}}=F_{4}^{2}\times F_{6}^{3}$, $F_{24}^{5^{00}}=F_{8}^{3}\times F_{3}^{2}$, $F_{24}^{6^{0}}=F_{4}^{3}\times F_{6}^{3}$, $F_{24}^{6^{00}}=F_{8}^{4}\times F_{3}^{2}$, $F_{48}^{6^{0}}=F_{8}^{3}\times F_{6}^{3}$, $F_{48}^{6^{00}}=F_{16}^{4}\times F_{3}^{2}$.

Table 2. Multiplication table of polytopes

Polytopes	Cube	Dodecahedron
Cube	$P_{64}^{6}(12F_{32}^{5})$	
Dodecahedron	$P_{160}^{6}(12F_{40}^{5},6F_{80}^{5})$	$P_{400}^{6}(24F_{100}^{5})$
Icosahedron	$P_{96}^{6}(20F_{24}^{5},6F_{48}^{5})$	$P_{240}^{6}(20F_{60}^{5^{0}},12F_{60}^{5^{00}})$
4-simplex	$P_{40}^{7}(5F_{32}^{6^{0}},6F_{20}^{6})$	$P_{160}^{7}(16F_{80}^{6},12F_{40}^{6})$
4-cube	$P_{128}^{7}(6F_{32}^{6^{0}},8F_{64}^{6})$	$P_{320}^{7}(8F_{160}^{6},12F_{80}^{6})$

$F_{32}^{6^{0}}=F_{4}^{3}\times F_{8}^{3}$, $F_{32}^{6^{00}}=F_{8}^{4}\times F_{4}^{2}$, $F_{60}^{5^{0}}=F_{3}^{2}\times F_{20}^{3}$, $F_{60}^{5^{00}}=F_{12}^{3}\times F_{5}^{2}$.

Table 3. Multiplication table of polytopes

Polytopes	Icosahedron	4-Simplex
Icosahedron	$P^6_{144}(40F^5_{36})$	
4-simplex	$P^7_{60}(5F^6_{48}, 20F^6_{15})$	$P^8_{25}(10F^7_{20})$
4-cube	$P^7_{192}(8F^6_{96}, 20F^6_{48})$	$P^8_{80}(5F^7_{64}, 8F^7_{40})$
4-cross-polytope	$P^7_{96}(16F^6_{48}, 20F^6_{24})$	$P^8_{40}(5F^7_{32}, 16F^7_{20})$

Table 4. Multiplication table of polytopes

Polytopes	4 –Cube	4 –Cross-Polytope
4 –cube	$P^8_{256}(16F^7_{128})$	
4 – cross-polytope	$P^8_{128}(16F^{7^0}_{64}, 8F^{7^{00}}_{64})$	$P^8_{64}(32F^7_{32})$

$F^{7^0}_{64} = F^3_4 \times F^4_{16}$, $F^{7^{00}}_{64} = F^4_8 \times F^3_8$.

a separate molecule, but a collection of molecules. This raises the problem of filling spaces with molecules, i.e. convex geometric figures of different dimensions. The addition of other molecules (indifferent or not) leads to the formation of a cluster. There are various patterns in the formation of clusters (Lord, Mackay & Ranganathan, 2006; Ilyushin, 2003). You can distinguish different levels of organization in clusters. From a geometric point of view, this amounts to a hierarchical filling of the space (Zhizhin, 2012). However, the hierarchical filling of space is always limited by the size of some area of space, primarily because when distance from the original molecule the distance between the particles (or points) increases significantly (Zhizhin, 2014). This inevitably leads to a weakening of the bonds between molecules and, as a consequence, to the detachment of molecules. Therefore, in order to further fill of the space, it is necessary to proceed to space translation a certain set of atoms or molecules (cluster). If a molecule in geometric relation is a multidimensional formation, then the problem arises of filling the space by means of translation by a multidimensional convex body. If a molecule or molecular formation is a multidimensional body, then this body must be translated from the coordinates of a multidimensional space. Moreover, in order for such a translation to be possible, the body itself must have a certain structure, i.e. to have selected directions along the coordinates in itself. Then translation of this body in the selected directions will not cause a violation of the body shape. The products of polytopes considered in this chapter just create forms in which there are selected directions along the coordinates of a multidimensional space. They are created by the very mechanism of the product polytopes. The mathematical apparatus developed in the chapter for determining the shape of a body that is a product of polytopes can be applied to determining the shape of bodies not listed in Tables 1 - 4. The diffraction pattern of the quasicrystal, shown in the Figure 1 in Chapter 4, reveals the nanostructure of the intermetallic compound $Al_{72}Ni_{20}Co_8$. You can see similar structures on other substances (Shechtman et al., 1984; Mukhopadhyay et al., 1993; Zhang & Kelton, 1993). The diffraction pattern clearly shows the system of parallel lines formed as result of the translation of molecules along different directions of coordinates (Shevchenko, Zhizhin & Mackay, 2013). At the same time, the lines formed as result of the hierarchical filling of space are visible. In fact, from each point one can see the formation of lines as a result of the hierarchical filling of space (if we neglect the influence of the experimental features on the diffraction pattern - the axial symmetry caused by the axial symmetry of the electron beam directed to the sample of matter). Just as there is an expansion of space at each point of space in the model of the expansion of the Universe (Silk, 1980).

The Chapter 7 will show how the existence of a system of parallel lines in a separate multidimensional molecular complex leads to the filling of a multidimensional space. It should be noted that the bases of all bodies of higher dimension that fill multidimensional spaces are the bodies of Plato.

REFERENCES

Alexandrov, P. S. (1975). *Introduction to homological dimension theory and general combinatorial topology*. Moscow: Science.

Fomenko, A. T. (1992). *Visual geometry and topology. Mathematical images in the real world*. Moscow: Moscow State University.

Ilyushin, G. D. (2003). *Modeling of self-organization processes in crystal-forming systems*. Moscow: URSS.

Lord, E. A., Mackay, A. L., & Ranganathan, S. (2006). *New geometry for new materials*. Cambridge, UK: Cambridge University Press.

Mukhopadhyay, N.K. (1993). Diffraction studies of icosahedral phases in $Al_{70}Fe_{20}W_{10}$. *Journal of Non-Crystalline Solids, 153-154*, 1193 – 1197.

Panina, G. Y. (2006). *The Algebra of Polyhedrons*. Academic Press.

Poincare A. (1895). Analysis situs .*J. de é Ecole Polytechnique, 1,* 1 – 121.

Pontryagin, L. S. (1976). *The foundations of combinatorial topology*. Moscow: Science.

Robertson, S. A. (1984). *Polytopes and Symmetry*. Cambridge, UK: Cambridge University Press.

Shechtman, D., Blech, I., Gratias, D., & Cahn, J. W. (1984). Metallic phase with longerange orientational order and no translational symmetry. *Physical Review Letters, 53*(20), 1951–1953. doi:10.1103/PhysRevLett.53.1951

Shevchenko, V. Ya., Zhizhin, G. V., & Mackay, A. (2013). On the structure of the quasi-crystals in the high dimension space. In M. V. Diudea (Ed.), *Diamonds and related nanostructures* (pp. 311–320). Dordrecht: Springer. doi:10.1007/978-94-007-6371-5_17

Silk, J. (1980). *The big bang. The creation and evolution of the Universe*. San Francisco: W.H. Freeman and Company.

Zang, X., & Kelton, K. (1993). High-order crystal approximant alloys $Ti_{54}Zr_{26}Ni_{20}$. *Journal of Non-Crystalline Solids, 153-154*, 114–118. doi:10.1016/0022-3093(93)90325-R

Zhizhin, G. V. (2012, October). *Hierarchical filling of spaces with polytopes*. Paper presented at "St. Petersburg Scientific Forum: Science and Human Progress". 7th St.- Petersburg meeting of Nobel Prize laureates, St. Petersburg, Russia.

Zhizhin, G. V. (2014). *World – 4D*. St. Petersburg: Polytechnic Service.

Zhizhin, G. V. (2015, November). *Polytopic prismahedrons – fundamental regions of the n-dimension nanostructures*. Paper presented at The International conference "Nanoscience in Chemistry, Physics, Biology and Mathematics", Cluj-Napoca, Romania.

Zhizhin, G.V. (2017). N-prismahedrons and their dual polytopes. *International Journal Chemical Modeling, 5*(4).

Ziegler, G. (1995). *Lectures on polytopes*. New York: Springer. doi:10.1007/978-1-4613-8431-1

KEY TERMS AND DEFINITIONS

5-Cross-Prism: The product of 4-cross-polytope by one dimension segment.

5-Simplex-Prism: The product of 4-simplex by one dimension segment.

6-Complex-Polytopic Prismahedrons: The product of tetrahedron by the tetrahedron.

6-Complex-Tetrahedral Prismahedrons and Octahedral Prismahedrons: The product octahedral by the tetrahedral.

6-Simplex-Polytopic Prismahedron: The product of 4-simplex by the triangle.

Cube-Polytopic Prismahedron: The product of cube by the triangle.

Dodeca-Polytopic Prismahedron: The product of dodecahedron by the triangle.

Dodecahedral Prism: The product of dodecahedral by one dimension segment.

Hierarchical and Translation Filling Spaces: At the same time hierarchical and translation filling of a multidimensional spaces.

Icosa-Polytopic Prismahedron: The product of icosahedron by the triangle.

Icosahedral Prism: The product of icosahedral by one dimension segment.

N*3-Angular Prismahedron: The product of n-angle by the triangle.

Octahedral Prism: The product of octahedral by one dimension segment.

Octahedral Prismahedron: The product of octahedral by the triangle.

Polytopic Prismahedron: The product of polytope by one dimension segment.

Tetrahedral Prism: The product of tetrahedral by one dimension segment.

Tetrahedral Prismahedron: The product of tetrahedral by the triangle.

Triangular Prismahedron: The product of triangle by the triangle.

Chapter 6
Polytopes Dual to Polytopic Prismahedrons

ABSTRACT

The polytopes are dual to polytypic prismahedrons. In particular, polytopes dual to the product of two canons. It is shown that these polytopes form a new class of polytopes with different values of the incidence of elements of low-dimensional polytopes to polytopes of higher dimension entering the polytope. If the polygons in their product have equal sides, then the dual polytope to the product consists of tetrahedrons, and the degree of incidence of the edge of the dual polytope is determined by the number of sides of the polygon. The existence of a previously unknown polytope consisting of one hundred tetrahedrons is established. Its election is constructed, all its constituent tetrahedrons are listed.

THE INCIDENCE IN POLYTOPE

In Chapter 4 it was established that one of the types of semi-regular polytopes, as deviations from the conditions for the correctness of polytopes that occur in the structures of chemical compounds, are poly-incident polytopes. In each of these polytopes there are simultaneously edges with different incidence values of elements of higher dimensional polytopes. We consider the question of the incidence of elements of polytopes in a more general form, i.e. let us consider the incidence of elements of polytopes of different dimensions to each other.

DOI: 10.4018/978-1-5225-4108-0.ch006

The incidence in polytopes indicates to what number of elements of higher dimension the given element of lower dimension belongs. Let's denote *e*(*d*) - an element of dimension *d*; k_{d_i,d_j} - the value of the incidence of the element with dimension d_i in relation to the elements of dimension $d_j (d_j > d_i)$. For regular polytopes because of their uniform values of the incidence k_{d_i,d_j} are constant for the whole polytope in all dimension range from 0 to *n* (*n*-dimension of the polytope). Obviously, that $k_{d_i,d_n} = 1$ for any $d_i < d_n$.

In a polygon we have $e(0) \in 2e(1); e(1) \in e(2)$.

In a polyhedron we have

$$e(3) : e(0) \in k_{d_0,d_1} e(1); k_{d_0,d_1} = 3, 4, 5; e(1) \in 2e(2); e(2) \in e(3).$$

In four-dimensional polytopes relations of the incidence have the following values.

In a simplex:

$$e(0) \in 4e(1); e(1) \in 3e(2); e(2) \in 2e(3); e(3) \in e(4); e(0) \in 4e(3); e(0) \in 4\,e(2).$$

In a hypercube:

$$e(0) \in 4e(1); e(1) \in 3e(2); e(2) \in 2e(3);$$
$$e(3) \in e(4); e(0) \in 3e(3); e(0) \in 6e(2); e(1) \in 3\,e(3).$$

In a 4-cross-polytope:

$$e(0) \in 6e(1); e(1) \in 4e(2); e(2) \in 2e(3); e(3) \in e(4);$$
$$e(0) \in 8e(3); e(0) \in 10e(2); e(1) \in 4e(3).$$

In semi-regular polytopes relations of incidence keep their form the same as in regular polytopes (Zhizhin, 2014). But there different figures in one polytope may serve as elements *e*(2), though all vertices of these semi-regular polytopes are superposed by motion. If in a polytope there are vertices which are not superposed by motion, then relations of incidence are variable in a polytope, for example, in a triangle prism. A prism can be considered a semi-regular polytope because it has two triangle faces and three of square faces. If we build a polyhedron dual to this prism, we'll receive an irregular polyhedron. Let's connect in a prism the centers of flat faces incident to one edge. We'll get a double pyramid (Figure 1).

In a double pyramid a two vertices are incident to three edges, and another two vertices are incident to four edges. The same picture will be if we take a pentagonal prism.

Two vertices of double pentagonal pyramid are incident to 5 edges, and 5 of the remaining vertices are incident to 4 edges (Figure 2).

Further it will be shown that the polytopes dual to polytopes products have not only vertices with different values of incidence to the edges, but the edges with different values of incidence to three-dimensional figures. This new type of polytopes will be called poly-incident polytopes (see Chapter 4).

POLYTOPE DUAL TO THE PRODUCT OF TWO TRIANGLES

In Chapter 5 the structural formula of the product of two triangles (1) was obtained, according to which this 4-polytope has 9 vertices and 6 triangular prisms act as facets. This polytope has 18 edges, 9 squares and 6 triangles as two-dimensional elements. In a dual polytope owing to inverse of inclusion relation (Grunbaum, 1969) 6 vertices, 15 edges, 18 two-dimensional elements, 9 three-dimensional facets shall be. Each three-dimensional face has the number of two-dimensional faces equal to double ratio of the number of

Figure 1. Triangle prism and double pyramid dual to it

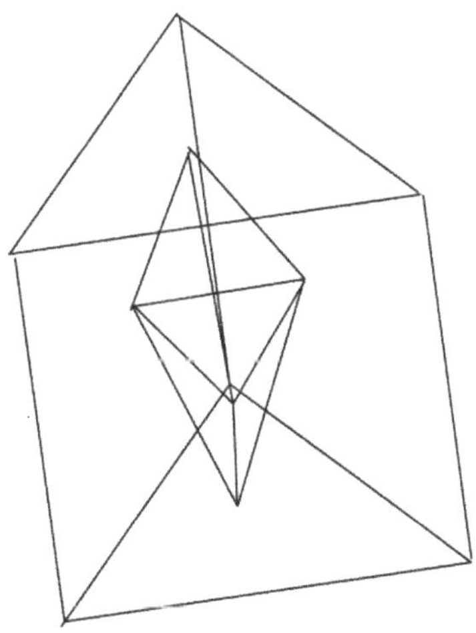

Figure 2. Pentagonal prism and a double pyramid dual to it

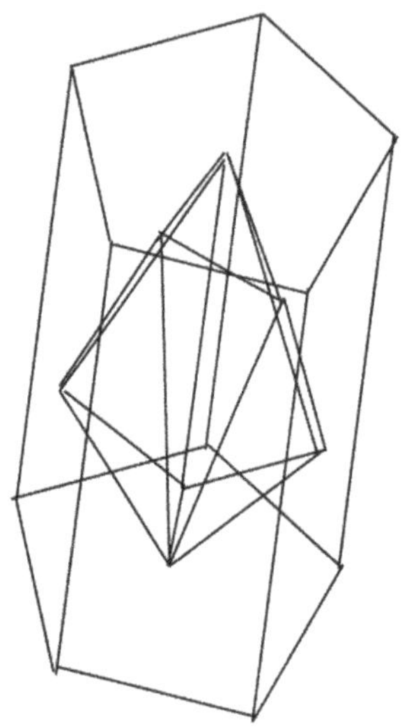

two-dimensional faces to the number of three-dimensional faces. Doubling happens due to the fact that each flat face in a 4-polytope simultaneously belongs to two three-dimensional faces, i.e. the number of flat faces belonging to three-dimensional faces of a dual polytope is $2 \cdot 18/9 = 4$. Therefore, the three-dimensional faces in a dual polytope are tetrahedrons, since only a tetrahedron has four flat faces. For the construction of a dual polytope let's use a sequence of actions that will be used further for construction of more complex polytopes. Let's denote vertices of the polytope in Chapter 5 by symbols $a_1,...,a_9$ (Figure 3).

Figure 3. Triangular prismahedron

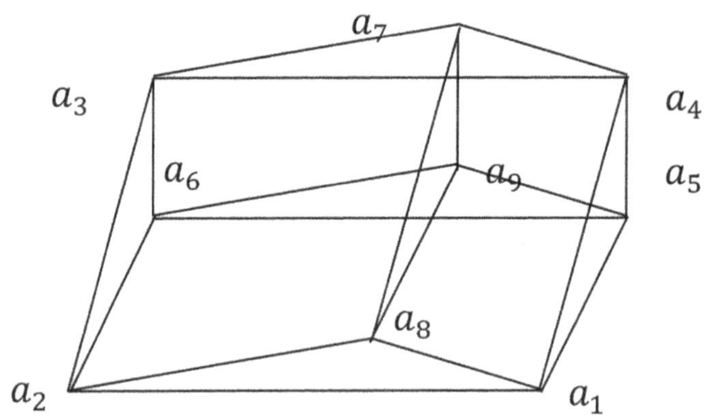

Let's enumerate triangular prismahedrons (facets) forming this polytope:

$$1)a_1a_2a_3a_7a_8a_9;2)a_2a_3a_4a_5a_8a_9;3)a_1a_2a_5a_6a_7a_8;$$
$$4)a_1a_2a_3a_5a_6a_4;5)a_1a_6a_3a_4a_7a_9;6)a_4a_6a_5a_7a_8a_9.$$

Since the edges of a dual polytope shall connect the centers of polytope facets adjoining each other, so, by introduced designations of facets, we define neighboring facets having common flat faces by introduced designations of facets. Therefore, just these facets shall be connected by edge in a dual polytope (Zhizhin, 2017). Executing this analysis, we'll come to the projection of a dual polytope on a two-dimensional plane (Figure 4).

In Figure 4 using the condition that each triangle shall be overlapped twice, we really find 9 tetrahedrons:

$$T_1 = 1246, T_2 = 2345, T_3 = 1245, T_4 = 6435, T_5 = 1635,$$
$$T_6 = 1362, T_7 = 1253, T_8 = 1645, T_9 = 6243.$$

Let's enumerate the edges and define their belonging to the listed tetrahedrons.

The edges: 61, 12, 24, 45, 35, 63, 13, 14, 34, 62, 25, 65, 15, 32, 64;

Figure 4. Polytope dual to triangular prismahedron

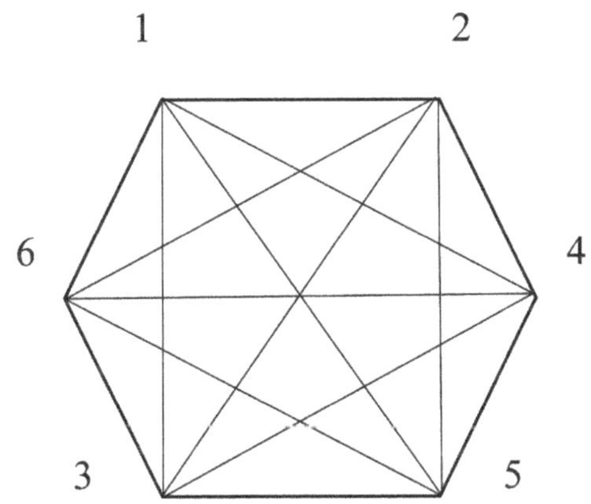

$61 \in T_1, T_6, T_5, T_8; 12 \in T_1, T_3, T_6, T_7; 24 \in T_1, T_2, T_3, T_9;$
$45 \in T_2, T_4, T_8, T_3; 35 \in T_2, T_4, T_5, T_7;$
$63 \in T_6, T_4, T_5, T_9; 13 \in T_7, T_5, T_6; 14 \in T_1, T_3, T_8;$
$34 \in T_2, T_4, T_9; 62 \in T_1, T_6, T_9; 25 \in T_2, T_3, T_7;$
$65 \in T_4, T_5, T_8; 15 \in T_3, T_5, T_7, T_9; 32 \in T_2, T_6, T_7, T_9; 64 \in T_1, T_4, T_8, T_9.$

Thus, there is a wonderful fact: the edges have different values of incidence to edges. The edges 13, 34, 14, 62, 25, 65 have value of incidence 3; the edges 61, 12, 24, 45, 35, 63, 15, 64, 32 have value of incidence 4. It occurs when the triangular prismahedron has all edges with equal values of incidence.

GENERAL ANALYSIS OF THE PRODUCT OF ANY TWO CONVEX POLYGONS

As a consequence of theorem 3 of Chapter 5 was received the structural formula (25) of the product of polytope P^n by by a polygon. If both factors in (25) of Chapter 5 are polygons, the formula (25) take the form of

$$P_m^2 \times Q_t^2 = P_{tm}^4(mF_{t2}^3, tF_{2m}^3). \tag{1}$$

According to Chapter 5, we'll call polytopes (1) *mt*-angular prismahedrons.

Theorem 1

Mt-angular prismahedron is a 4-polytope consisting of *m* *t*-angular prisms and *t* *m*-angular prisms.

Proof

As a result of the product of *m*-angular polygon by *t*-angular polygon, according to the definition of polytopes product (Ziegler, 1995), is a polytope with facets forms. These facets are the products of each side of *m*-angular polygon by *t*-angular polygon and, accordingly, the products of each sides of *t*-angular polygon by *m*-angular polygon. The first from the products give *m* *t*- angular prisms, the second ones – give *t* *m*-angular prisms. Let's check by Euler-Pioncare's equation (2) in Chapter 1 (Pioncare, 1880) that the polytopes formed as a result of the product of polygons satisfy this equation for polytope dimension $n = 4$. In this case the number of elements of dimension 0 (vertices) owing to constructing polytopes is equal to *mt.* The number of one-dimension elements (edges) is composed of the number of edges issuing from all vertices with account of each edge passing through two vertices. Since 4 edges issue from each vertex owing to polygons product, the total number of edges is $\frac{mt4}{2} = 2mt$. The number of two-dimensional faces is composed of numbers *m* and *t,* correspondingly *t*-angular and *m*-angular bases of prisms and 4-angular faces of prisms *mt.*

Substituting the received values of numbers of elements of different dimension in Euler-Pioncare's equation

mt – 2mt +(m+t+mt) – (m +t)=0.

Thus, Euler-Pioncare`s equation is satisfied, which proves theorem 1.

Consequence

Owing to duality in a polytope dual to *mt*-angular prismahedron, the number of vertices is equal to *m+t,* the number of edges is equal to *m+t+mt,* the number of two-dimensional faces is equal to 2*mt,* the number of three-dimensional faces is equal to *mt.*

Each three-dimensional face has the number of two-dimensional faces equal to double ratio $\frac{2mt}{mt}$, i.e. 4. Thus, in the general case of the product of any two convex polytopes the tree-dimensional figure is a tetrahedron, since only a tetrahedron, as a three-dimensional figure, has 4 two-dimensional faces.

4*3-ANGULAR PRISMAHEDRON AND ITS DUAL POLYTOPE

In Chapter 5 the image of 4*3-angular prismahedron (Figure 4) was received. Let's denote vertices of this polytope $a_1,...,a_{12}$ (Figure 5).

4*3-angular prismahedron includes three-dimensional figures: triangular prisms

*Figure 5. 4*3-angular prismahedron*

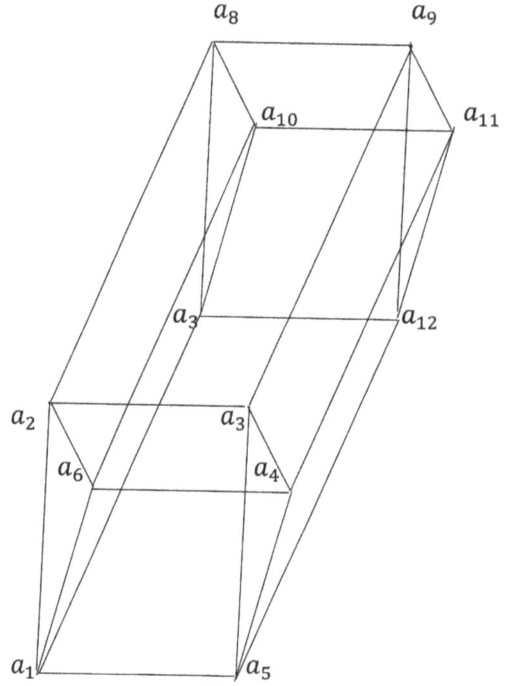

$1)a_1a_2a_6a_7a_8a_{10}, 2)a_3a_4a_5a_9a_{11}a_{12}, 3)a_1a_2a_6a_3a_4a_5, 4)a_7a_8a_{10}a_{11}a_{12};$

tetragonal prisms

$5)a_2a_6a_3a_4a_8a_{10}a_9a_{11}, 6)a_1a_6a_4a_5a_7a_{10}a_{11}a_{12}, 7)a_1a_2a_3a_5a_7a_8a_9a_{12}.$

For constructing dual polytope we define contacts of the prisms by equal sets of symbols a_i. In the result we receive a projection of dual polytope (Figure 6).

In polytope in Figure 6 one can single out 12 tetrahedrons with flat faces overlapped two times: 2564, 4527, 2537, 2563, 4561, 4571, 5731, 2746, 1536, 3627, 6714, 7136. This polytope have 19 edges: 17, 16, 15, 14, 13, 37, 36, 35, 32, 27, 26, 24, 25, 57, 56, 54, 47, 46, 67. From them edges 17, 14, 32, 24 have incidence value 3 with respect to tetrahedrons and the remaining 15 edges have incidence value 4. The vertices of dual polytope 3, 2, 1, 4 are incident to 5 edges and vertices 5, 6, 7 are incident to 6 edges.

*Figure 6. Polytope dual to 4*3-angular prismahedron*

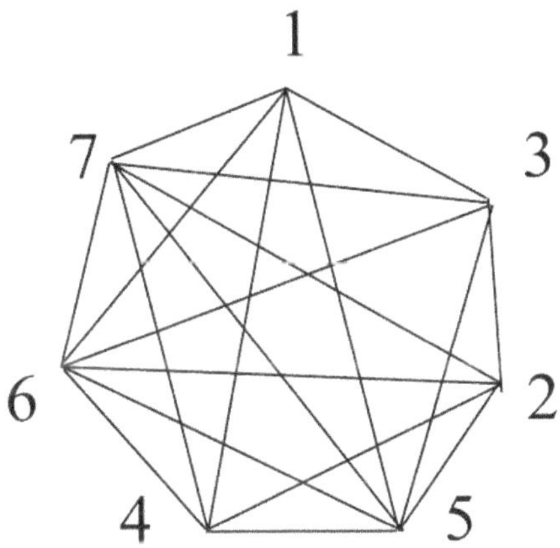

3*6-ANGULAR PRISMSHEDRON AND A POLYTOPE DUAL TO IT

In Figure 7 a projection of 3*6-angular prismahedron with marked 18 vertices is shown.

It has three 6-angular prisms

$1)c_1c_2c_3c_4c_5c_6c_7c_8c_9c_{10}c_{11}c_{12}, 2)c_7c_8c_9c_{10}c_{11}c_{12}c_{13}c_{14}c_{15}c_{16}c_{17}c_{18},$

$1)c_1c_2c_3c_4c_5c_6c_{13}c_{14}c_{15}c_{16}c_{17}c_{18},$

and six 3-angular prisms

$4)c_1c_7c_6c_{13}c_{12}c_{18}, \ 5)c_1, c_2, c_7, c_{13}, c_{14}, c_8,$

$6)c_2c_3c_8c_9c_{14}c_{15}, 7)c_3c_4c_9c_{10}c_{15}c_{16}, 8)c_4c_5c_{16}c_{10}c_{11}c_{17}, 9)c_5c_6c_{11}c_{12}c_{17}c_{18}.$

Analyzing contacts of the prisms on coinciding sets of symbols, we construct a projection of the polytope dual to 6 * 3-angular prismahedron (Figure 8).

In polytope in Figure 8 we can single out 18 tetrahedrons:

$T_1 = 1234, T_2 = 1235, T_3 = 1236,$

*Figure 7. 3*6-angular prismahedron*

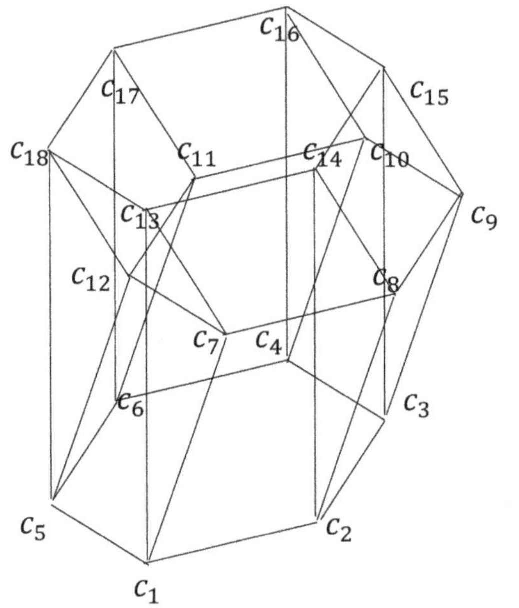

*Figure 8. Polytope dual to 3*6-angular prismahedron*

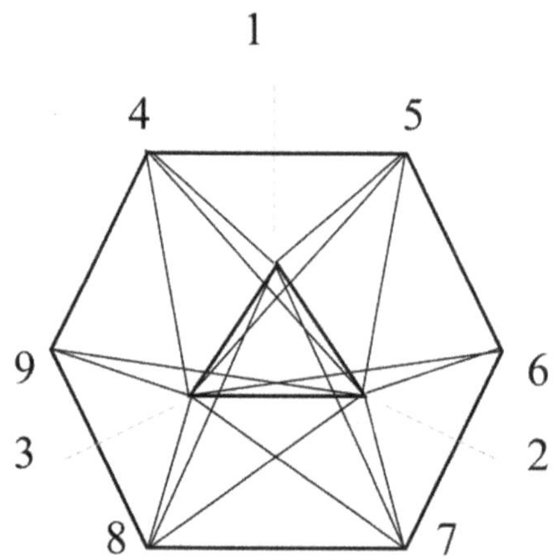

$T_4 = 1237, T_5 = 1238, T_6 = 1239, T_{10} = 3289, T_{11} = 1394, T_{12} = 1398,$

$T_7 = 1265, T_8 = 1267, T_9 = 2378,$

$T_{13} = 4325, T_{14} = 4315, T_{15} = 4215, T_{16} = 3267, T_{17} = 1378, T_{18} = 1278.$

The polytope has 27 edges with different values of incidence to tetrahedrons: edge

$45 \in T_{13}, T_{14}, T_{15}; 12 \in T_1, T_8, T_{15}, T_{18}; 15 \in T_2, T_7, T_{14}, T_{15};$
$14 \in T_1, T_{11}, T_{14}, T_{15}; 16 \in T_3, T_7, T_8;$

$56 \in T_7; 19 \in T_6, T_{11}, T_{12}; 94 \in T_{11}; 13 \in T_1, T_6, T_{11}, T_{12}, T_{14}, T_{17};$
$42 \in T_1, T_{13}, T_{15}; 35 \in T_2, T_{13}, T_{14};$

$43 \in T_1, T_{11}, T_{13}, T_{14}; 32 \in T_1, T_6, T_9, T_{10}, T_{13}, T_{16};$
$93 \in T_6, T_{10}, T_{12}; 26 \in T_3, T_7, T_8, T_{16}; 92 \in T_6, T_{10};$

$36 \in T_3, T_{16}; 17 \in T_4, T_8, T_{17}, T_{18};$
$52 \in T_2, T_7, T_{13}, T_{14}; 87 \in T_9, T_{17}, T_{18}; 18 \in T_5, T_{12}, T_{17};$

$37 \in T_4, T_9, T_{16}, T_{17}; 82 \in T_5, T_9, T_{10}, T_{18};$
$27 \in T_4, T_8, T_9, T_{16}, T_{18}; 83 \in T_5, T_9, T_{10}, T_{12}, T_{17};$

$76 \in T_8, T_{16}; 98 \in T_{10}, T_{12}.$

Thus, edges 56, 94 have incidence value equal to 1; edges 92, 36, 76 have incidence value equal to 2; edges 45, 16, 19, 42, 35, 87 have incidence value equal to 3; edges 15, 14, 43, 93, 26, 17, 52, 82 have incidence value equal to 4; edges 27, 83 have incidence value equal to 5; edges 12, 13, 32 have incidence value equal to 10.

In addition, we note that vertices 1, 2, 3 are incident to 14 tetrahedrons; vertices 4, 5 are incident to 5 tetrahedrons; vertices 6, 9 are incident to 4 tetrahedrons; vertices 7, 8 are incident to 6 tetrahedrons.

This method can be used for construction of dual polytopes to many discussed in this paper, the polytopes with simultaneous use of two-dimensional faces of different shapes. All these polytopes are poly-incident, i.e. they have edges (and vertices) with different values of incidence.

N-PRISMAHEDRONS AND POLYTOPES DUAL TO THEM

Let's call n-prismahedron a polytope of dimension 4, with facets in the form of prisms with n - angular bases. In item 4*3-ANGULAR PRISMAHEDRON AND ITS DUAL POLYTOPE a 3-prismahedron was considered. Now let'us turn to the study of n - prismahedrons for any n.

Theorem 2

The product of two n-angular polygons is an n-prismahedron with the number of facets $2n$.

Theorem 2 follows immediately from theorem 1 when $m = t = n$.

Corollary

In a polytope dual to n-prismahedron the number of vertices is equal to $2n$, the number of edges is equal to $2n + n^2$, the number of two-dimensional faces is equal to $2\ n^2$, the number of three-dimensional faces is equal to n^2. This follows directly from the corollary to theorem 1.

Theorem 3

Polytopes dual to n-prismahedrons are those composed of tetrahedrons. For $n = 4$ the dual polytope is a 4-cross-polytope, for $n = 10$ the dual polytope is a previously unknown polytope consisting of 100 tetrahedrons (100-cell one) with a degree of incidence equal to 5, for other values of $n \geq 3$ the dual are the poly-incident polytopes with a medium degree of edge incidence tending to 6 with an infinite increase of n.

Proof

As polytopes dual to n-prismhedrons have the number of vertices equal to $2n$, owing to the inverse of inclusion relation, the number of edges is equal to $2n + n^2$, the number of 2-faces is equal to $2n^2$, the number of 3-faces is equal to n^2 Then each 3-figure has the number of 2-faces $2\frac{2n^2}{n^2} = 4$ $2\frac{2n^2}{n^2} = 4$ (multiplication by 2 is due to the fact that each 2-face belongs at once to two 3-shapes). Thus, a 3-figure is a tetrahedron, since only tetrahedron, as a

3-figure, has 4 2-faces. As everybody knows, the number of edges in a tetrahedron is 6. We can write this number on the basis of set numbers of different dimension in a dual polytope: $6 = k\frac{2n + n^2}{n^2}$, k is the degree of edge incidence (the edge belongs to k-tetrahedrons) under the assumption that all the edges have this degree of incidence, the same for all. Hence, we find the degree of incidence

$$k = \frac{6n}{n + 2}, n \geq 3. \tag{2}$$

Equation (2) in integers has two solutions: $k = 4$ when $n = 4$ and $k = 5$ when $n = 10$. For the other possible values of n, according to (2) the degree of incidence of edges is a fractional value. This means that in these cases in a dual polytope edges with varying degrees of incidence present simultaneously, i.e. it is a poly-incident polytope.

Figure 3, Figure 9, Figure 10 give examples of 3-prismahedron, 5-prismahedron, 6-prismahedron respectively.

The known hypercube is 4-prismahedron.

For $n = 4$ in a dual polytope the number of vertices is $2n = 8$, the number of edges is $2n + n^2 = 24$, the number of $2D$-faces is $2n^2 = 32$, the number of $3D$-faces is $n^2 = 16$. This is a 4-cross-polytope dual to a hypercube. Each edge of this polytope has a degree of incidence 4.

For $n = 5$ in a dual polytope the number of vertices is $2n = 10$, the number of edges is $2n + n^2 = 35$, the number of $2D$-faces is $2n^2 = 50$, the number of $3D$-faces is $n^2 = 25$. In the projection on the plane this polytope is presented in Figure 11.

In Figure 11, tetrahedrons are: 1)*abgh, 2)dckh, 3)kcnd, 4)efnd, 5)agfe, 6) hkgn, 7)knfh, 8)fngk, 9)gfhn, 10)ghfk, 11)ahfg, 12)gbkh, 13)hcnk, 14)fkdn, 15)egnf, 16)abec, 17)bcad, 18)cdeb, 19)edac, 20)aebd, 21)agbe, 22)bhac, 23)kcbd, 24)ndec, 25)efad.* Every vertice of the polytope belongs to 10 tetrahedrons. Every $2D$-face belongs to two tetrahedrons (also as for $n = 3; 4$). From 35 edges 10 edges are incident to 3 tetrahedrons $gb \in$ 1), 12), 21); $bk \in$ 2), 12), 23); $hc \in$ 2), 13), 22); $nc \in$ 3), 13), 24); $kd \in$ 3), 14), 23); $fd \in$ 4), 14), 25); $en \in$ 4), 15), 24); $eg \in$ 5), 15), 21); $af \in$ 5), 11), 25); $ah \in$ 1), 11), 22). From 35 edges 15 edges are incident to 4 tetrahedrons $ag \in$ 1), 5), 11), 21); $bh \in$ 1), 2), 12), 22); $kc \in$ 2), 3), 13), 23); $nd \in$ 3), 4), 14), 24); $ef \in$ 4), 5), 15), 25); $eb \in$ 16), 18), 20), 21); $bd \in$ 17), 18), 20), 23); $ac \in$ 16),

Figure 9. 5-prismahedron

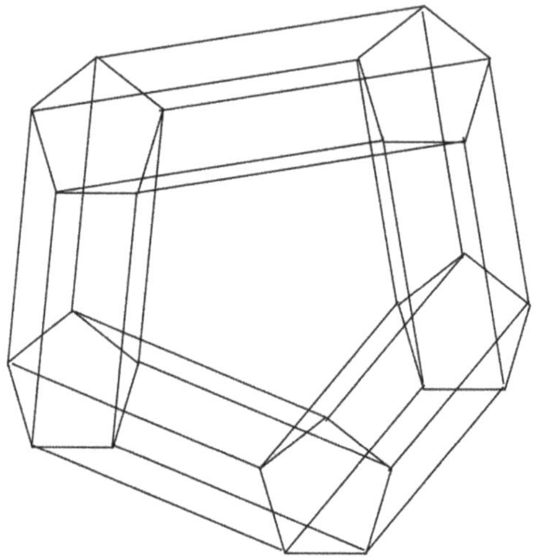

Figure 10. 6-prismahedron

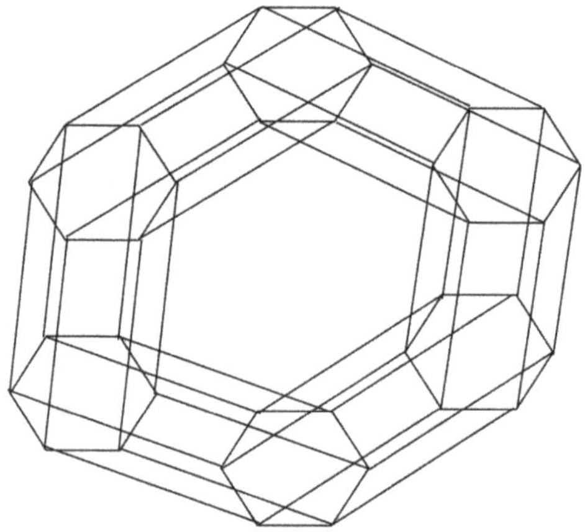

17), 19), 22); *ec* ∈ 16), 18), 19), 24); *ad* ∈ 17), 19), 20), 25); *hf* ∈ 7), 9), 10), 11); *hn* ∈ 6), 7), 9), 13); *kg* ∈ 6), 8), 10), 12); *kf* ∈ 7), 8), 10), 14); *gn ∈ 6), 8), 9), 15).* From 35 edges 10 edges are incident to 6 tetrahedrons: *ab* ∈ 1), 16), 17), 20), 21), 22); *bc* ∈ 2), 16), 17), 18), 22),23); *cd* ∈ 3), 17), 18), 19), 23), 24); *ed* ∈ 4), 8), 19), 20), 24), 25); *ae* ∈ 5), 16), 19), 20), 21), 25); *gh* ∈ 1), 6), 9), 10), 11), 12); *hk* ∈ 2), 6), 7), 10), 12), 13); *kn* ∈ 3), 6), 7), 8), 13), 14); *fn* ∈ 4), 7), 8), 13), 14); *gf* ∈ 5), 8), 9), 10), 11), 15). Thus, the average

Figure 11. Polytope dual to 5-prismahedron

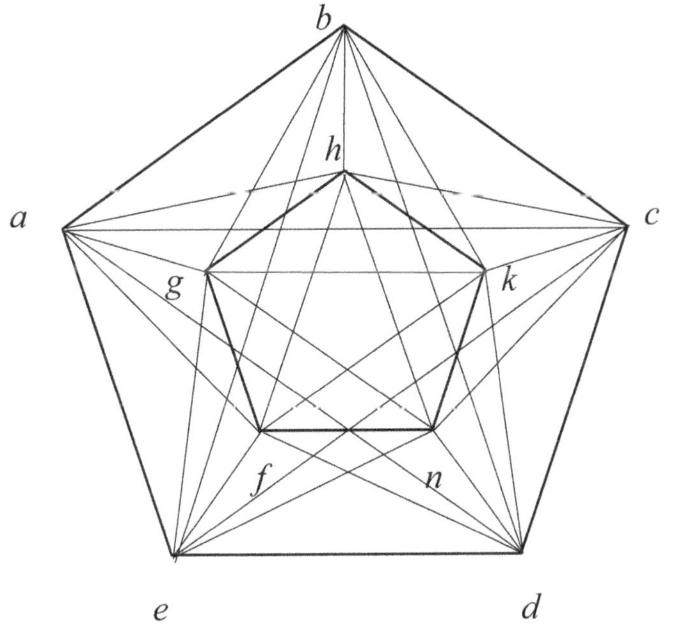

value of incidence degree is equal to: $\bar{k} = \frac{10 \cdot 6 + 15 \cdot 4 + 10 \cdot 3}{35} = \frac{30}{7}$ and coincides exactly with the degree of incidence according to equation (2).

By increasing *n*, the structure of a dual polytope gets greatly more complicated and there are difficulties in its constructing. For $n = 6$ in a dual polytope the number of vertices is $2n = 12$, the number of edges is $2n + n^2 = 48$, the number of 2*D*-faces is $2n^2 = 72$, the number of 3*D*-faces is $n^2 = 36$. This polytope is presented in Figure 12 in the projection on the plane.

In Figure 12 tetrahedrons are: 1) *hmng,* 2) *acdf,* 3) *hkmg,* 4) *abcf,* 5) *hkmn,* 6) *abcd,* 7) *hkno,* 8) *abde,* 9) *gmno,* 10) *fcde,* 11) *kmno,* 12) *bcde,* 13) *gkmo,* 14) *fbce,* 15) *ghko,* 16) *fabe,* 17) *ghno,* 18) *fade,* 19) *hkmb,* 20) *abck,* 21) *kmnc,* 22) *bdcm,* 23) *mnod,* 24) *cden,* 25) *gone,* 26) *fedo,* 27) *hgof,* 28) *afeg,* 29) *ghka,* 30) *fabh,* 31) *kmbc,* 32) *mncd,* 33) *oned,* 34) *gofe,* 35) *hgaf,* 36) *hkab.* From 48 edges each of 12 edges is incident to 7 tetrahedrons from the following list: *ab, bc, cd, de, fe, af, hk, km, mn, no, og, hg;* each of 24 edges is incident to 4 tetrahedrons: *ko, gm,hn, ho, kg, mo, kb, oe, fc, be, ad, ac, fd, bd, mc, gf, hm, gn, kn, ae, bf, ce, nd, ha;* each of 12 edges is incident to 3 tetrahedrons: *bm, kc, ge, fo, md, cn, ga, hf, ne,od, hb, ak.* Thus, the average

Figure 12. Polytope dual to 6-prismahedron

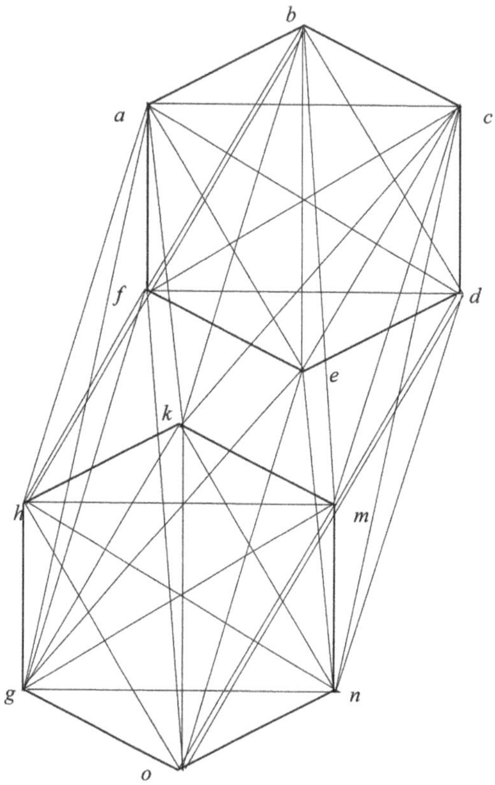

value of edges incidence degree is equal to $\bar{k} = \dfrac{12 \cdot 7 + 24 \cdot 4 + 3 \cdot 12}{48} = 4{,}5$, which fits exactly with the degree of incidence *k* according to equation (2).

Every vertex of this polytope is incident to 12 tetrahedrons. From 72 triangles, which are faces of tetrahedrons, some have degree of incidence 3 and the rest have degree of incidence 2.

The construction of a dual polytope with $n = 10$ is particularly interesting, since according to equation (2) in this case the degree of incidence of edges is the same for the whole polytope and equals to 5. For $n = 10$ in a dual polytope, the number of vertices is $2n = 20$, the number of edges is $2n + n^2 = 120$, the number of 2*D*-faces is $2n^2 = 200$, the number of 3*D*-faces is $n^2 = 100$. This polytope is presented in Figure 13 in the projection on the plane.

Tetrahedrons in Figure 13 are:

Figure 13. Polytope dual to 10-prismahedron

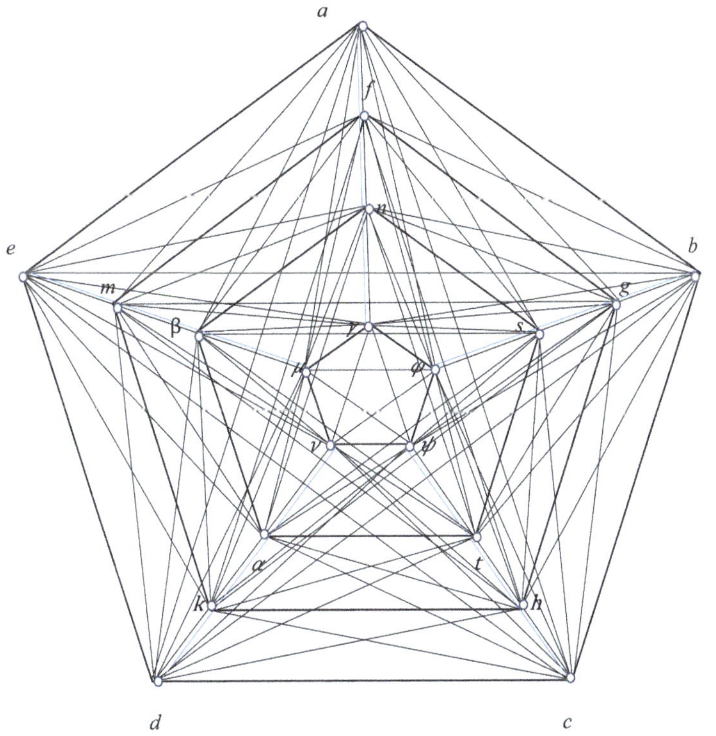

1) *dkhc,* 2) *cgbh,* 3) *fagb,* 4) *aefm,* 5) *demk,*

6) *ktha,* 7) t*hsg,* 8) n*fsg,* 9) m*fnβ,* 10) *β*m*kα,*

11) *α*t*ψv,* 12) *φψ*st, 13) *γφn*s, 14) *μγnβ,* 15) *βμαv,*

16) *φψγμ,* 17) *φψμγ,* 18) *φψγv,* 19) *γψvφ,* 20) *φγvμ,*

21) *d*hbc, 22) *cgab,* 23) *bfae,* 24) daem, 25) ekcd,

26) ktg*h,* 27) hsfg, 28) mgn*f,* 29) fkm*β,* 30) mhk*α,*

31) dc*vψ,* 32) *φ*c*bψ,* 33) a*γ*b*φ,* 34) *μ*ae*γ,* 35) *v*de*μ,*

36) *vdψμ,* 37) *cvφψ,* 38) *vφψb,* 39) a*φγμ,* 40) e*γvμ,*

41) φγνμ, 42) φνtψ, 43) φγψs, 44) μnφγ, 45) βγνμ,

46) kdth, 47) *hcsg,* 48) *gbnf,* 49) afmβ, 50) αemk,

51) α*k*hc, 52) *thgb,* 53) as*gf,* 54) fn*em,* 55) md*k*β,

56) *k*α*t*ψ, 57) t*h*φ*s,* 58) sgγ*n,* 59) nfμ β, 60) ανmβ,

61) t*h*αν, 62) ψ*sgt,* 63) snfφ, 64) nβγμ, 65) *k*βμα,

66) ktαβ, 67) *thsα*, 68) *sgnt,* 69) fn*mg,* 70) mnβα,

71) tανβ, 72) tsαψ, 73) *snt*φ, 74) *ns*βγ, 75) βαμ*n,*

76) m*h*αν, 77) tkgψ, 78) sf*h*φ, 79) γn*mg,* 80) kf βμ,

81) decν, 82) *cd*bψ, 83) *tac*φ, 84) *eb*γ*a,* 85) *da*μ*e,*

86) dm*h*ν, 87) ckgψ, 88) bfhφ, 89) *a*mgγ, 90) *ef*kμ,

91) *kc*αψ, 92) *thb*φ, 93) *sg*γ*a,* 94) e*nf*μ, 95) mνβ*d,*

96) kμ*e*α, 97) maβγ, 98) fnbφ, 99) fgψc, 100) thdν.

In t*he given* abo*ve list* of t*etrahe*drons *each* line *corre*sponds *to a* specific *type* of te*trahedr*on pr*ojecti*on, w*hich d*iffer *only b*y angle of rotation relative to the polytope projection center divisible by 72 degrees. The structure of a 100-cell polytope is quite complicated, but it is defined by enumerating all the tetrahedrons. If all the edges of a polytope have the same degree of incidence, the degree of incidence of vertices and triangles, which are faces of tetrahedrons, is variable.

By increasing the value of n in equation (2), the average degree of edges incidence tends to its maximum value of 6.

Theorem 3 is proved.

Note. It should be *n*oted that the number of elements of different dimension of the polytope in Figure 7 is 6 times smaller than th*e cor*responding values of a regular polytope consisting of 600 tetrahedrons (600-cell polytope, Coxeter, 1963) with the same degree of edges incidence. Since the proofs for the existence of 600-cell polytope in the works of Stringham (Stringham,

1880) and Coxeter (Coxeter, 1963) contain incorrectness (Zhizhin, 2014), there is no specific enumeration of tetrahedrons and their images do not carry specific information, the structure of the proposed 600-cell polytope remains unknown.

In Chapter 4 polytopes with different incidence values of edges to three-dimensional figures included in the polytope were considered. Graphs of these polytopes were constructed, with each three-dimensional body being taken for the top of the graph. By construction, these graphs are topologically equivalent to polytopes, dual to poly-incident polytopes. It is now clear that polytopic prizmahedrons, which play an important role in filling the space of higher dimension (Chapter 7), are polytopes dual to poly-incident polytopes. In Chapter 4, a poly-incident polytope with incidence values of edges 4 and 5 was not constructed. Only the graph of this polytope was constructed (Figure 14, Chapter 4). In this chapter, along with other prismahedrons of higher dimension, this polytope is constructed (Figure 11).

REFERENCES

Coxeter, H.S.M. (1963). Regular Polytopes. New York: John Wile & Sons, Inc.

Grunbaum, B. (1967). Convex Polytopes. London: Springer.

Poincare, *A. (1895). Analysis* situs. J. de é Ecole Polytechnique, 1, 1 – 121.

St*ringham, W. I. (*1880). Regular figures in n-dimensional space. American J*ournal of Mathematics, 3(1), 1*–14. doi:10.2307/2369441

Zhizhin, G. V. (2014). World – 4D. St. Petersbur*g: Polytechnic Service.*

*Zhiz*hin, G.V. (2017). N-prismahedrons and their dual polytope*s. Interna*tional Journal Chemical Modeling, 5(4).

Ziegler, G. (1995). Lectures on polytopes. New York: Springe*r. doi:10.1007/978-1-4613-8431-1*

KEY TERMS AND DEFINITIONS

Duality in Polytopes: If the facets of one polytope correspond to the vertices of another polytope, and the edges of one polytope correspond to common elements of the facets of another polytope, then these polytopes are dual.

Incidence in Polytopes: Incidence in polytopes define the number of elements of higher dimension the given element of lower dimension belongs.

Polyincident Polytopes: Polytopes in which elements of lower dimension have different incidence values for elements of higher dimension. Polytopes that are dual to polytopes products are polyincident polytopes.

Chapter 7
Stereohedrons and Partition of *n*-Dimensional Space

ABSTRACT

The process of hierarchical filling of space by p-dimensional regular polytopes is considered under the condition of large-scale discrete increase in the size of polytopes and preservation of their shape (scaling process). It is shown that the polytopic prismahedrons are a concrete realization of the stereohedrons. The polytopic prismahedrons have the necessary properties for translational filling of spaces of higher dimension without slits face to face. Moreover, it is proved that the polytopic prismahedrons forming such fillings can have common elements of any dimension included in the polytope. On the basis of the research carried out in spaces of higher dimension, a new paradigm for describing a discrete world has been put forward.

THE SCALING PROCESS AND HIERARCHICAL FILLING OF THE *N*-DIMENSIONAL SPACE

The problem of completing space by polyhedrons is one of the fundamental problems of mathematics, which has long attracted the attention of scientists. In 1900, D. Gilbert formulated 23 mathematical problems that require solution (Gilbert, 1901). One of these problems (eighteenth) was devoted to this question. It was formulated as follows: "Construction of space from congruent polyhedrons". This problem is especially complicated in the case of n-dimensional spaces (Delone, 1969), and up to the end it has not been solved

DOI: 10.4018/978-1-5225-4108-0.ch007

to this day under these conditions. The discovery of the processes of scaling occurring in natural media substantially enriches the problem of filling spaces with polyhedrons. The idea of scaling was advanced by Kadanov in 1966 when analyzing the processes of a second-order phase transition (Kadanov, 1966). According to this idea, the elementary cells of the high-symmetry phase before the phase transition in the process combine with the formation of enlarged elementary cells of the low-symmetry phase. The existence of this process it was confirmed experimentally. The idea of scaling allowed Wilson (Wilson, 1971) and Fisher (Fisher, 1972) to describe the second-order phase transition by the system of Ginzburg-Landau differential equations and to investigate it (Zhizhin, 2014a, 2014b, 2014c). The idea of scaling was used in analyzing processes in other fields of physics, for example, in analyzing the growth (enlargement) of clusters (Krapivsky, Redner & Ben-Naim, 2010). In the diffraction patterns of quasi-crystals of various intermetallic compounds (Abe, Yan & Pennycook, 2004; Munkhopadhyay at el., 1993; Zhang & Kelton, 1993), one can also see the enlargement of the shape of the figures that unite the group of glowing points of the diffraction patterns (see Chapter 4, Figure 1). This enlargement includes an increasing number of these points. Moreover, this process has no limit both in the large and in the smaller side. A certain model of the process of enlargement of objects can be a similar increase in the object's odds. This leads to representations about the hierarchical filling of space with some initial figure. The process of increasing the size of a figure is discrete and it is determined by the coefficient of geometric progression, the value of which depends on the shape of the figure. The concept of a growing geometric manifold was introduced and the hierarchical filling of the plane by various regular polygons (Zhizhin, 2010) was investigated, as well as the hierarchical filling of three-dimensional spaces by regular convex polyhedrons (Zhizhin, 2012; Zhizhin, 2014c).

Now, applying the methods developed in previous studies (Zhizhin, 2010, 2014c), we consider the hierarchical filling of a multidimensional space in the process of enlargement of multidimensional convex bodies. As was shown earlier when studying the hierarchical filling of a plane with polygons (Zhizhin, 2010), there can be various ways of filling it. In a multidimensional space are different ways of hierarchical filling with polytopes also. In Chapter 4, the method of hierarchical filling of space by polytopes was considered, based on the transition of a polytope to a dual polytope at each step of filling the space. In this case, investigating scaling processes, we need to consider a method for filling a space of higher dimension with polytopes, at which only the scale of the figure would change at each step, and the polytope's background would be preserved.

On Figure 1 presented projection of 4-simplex (Stringham, 1880).

Continuing the outer edges of the projection to the intersection with each other, connecting the resulting vertices with edges, we obtain an enlarged projection of the 4-simplex (Figure 2).

Obviously, the process of extending the 4-simplex can be continued further. Each time the size (length) of the edges of the next 4-simplex from the size of the edges of the previous 4-simplex is increased by $1 + \tau$ times. Where τ is the golden section. Since the coefficient of the geometric progression of the discrete increase of each side of the regular pentagon in the hierarchical filling of the pentagon plane is $1 + \tau$ (Zhizhin, 2010), then each segment joining the vertices of the pentagon increases in the same progression. The difference in the lengths of the segments of the projection of a regular 4-simplex into a two-dimensional plane is related to the design process, but their images in four-dimensional space have the same length. Thus, the geometric progression coefficient of the hierarchical filling of the space by a 4-simplex is $1 + \tau$.

On Figure 3 presented known projection of 4-cube (Stringham, 1880).

We represent the projection of the polytope 4-cube in a more symmetrical form (Figure 4).

Continuing the outer edges of the projection to the intersection with each other, connecting the resulting vertices with edges, we obtain an enlarged projection of the 4-cube (Figure 5).

Figure 1. The regular 4-simplex

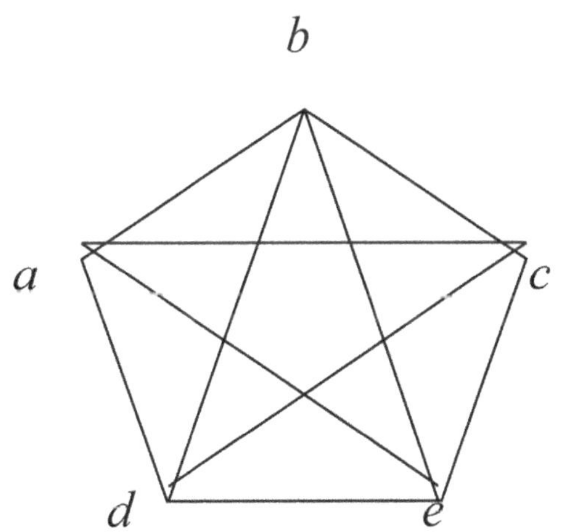

Figure 2. The hierarchical filling of the space by a 4-simplex

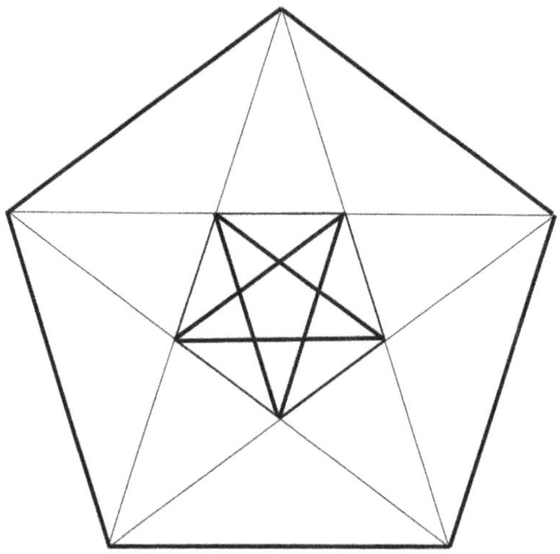

Figure 3. The regular 4-cube

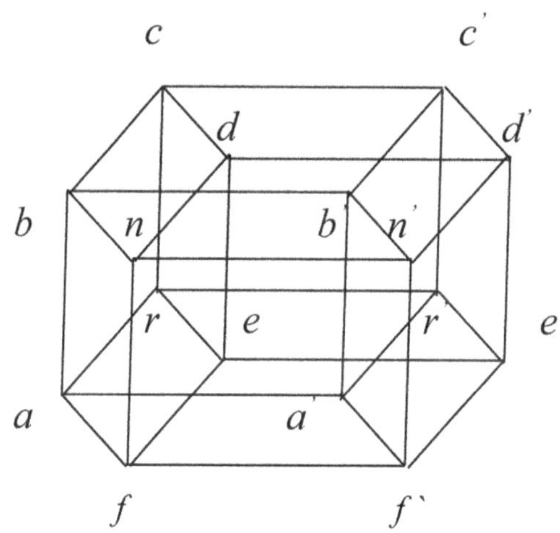

Obviously, the process of extending the 4-cube can be continued further. Easy to see that each time the size (length) of the edges of the next 4-cube from the size of the edges of the previous 4-cube is increased by $1+\sqrt{2}$ times. Thus, the geometric progression coefficient of the hierarchical filling of the space by a 4-cube is $1+\sqrt{2}$.

On Figure 6 presented known projection of 4-cross-polytope (Stringham, 1880).

Figure 4. The symmetrical form of polytope 4-cube

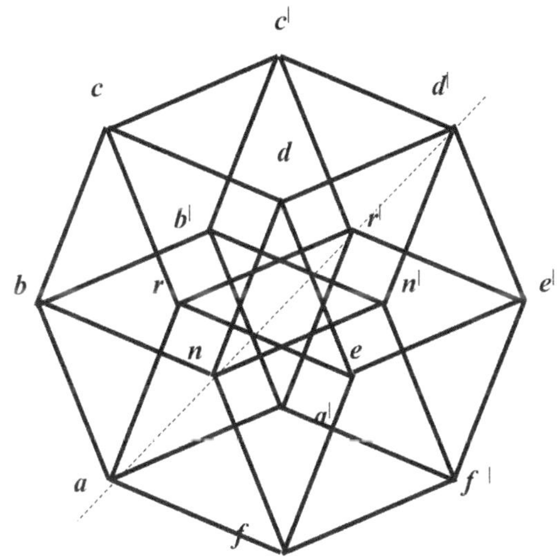

Figure 5. The hierarchical filling of the space by a 4-cube

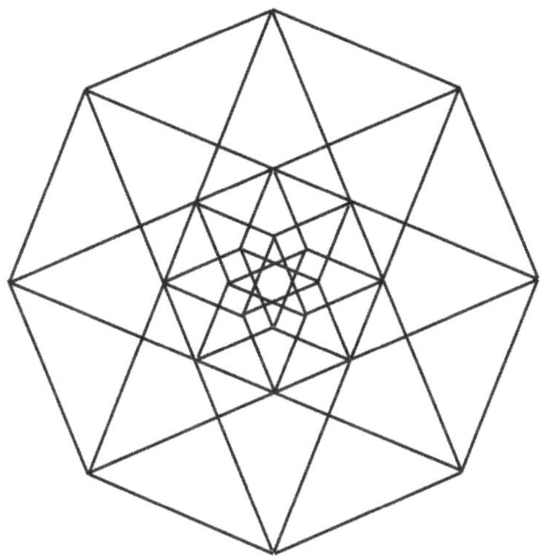

We represent the projection of the 4-cross-polytope in a more symmetrical form (Figure 7).

In this case, the continuation of the exterior edges of the projection gives an opportunity to obtain new vertices (there is no intersection of the edges). So we apply the second method of hierarchical filling, which was used to study the hierarchical filling of the plane (Zhizhin, 2010). We draw through

Figure 6. The regular 4-cross-polytope

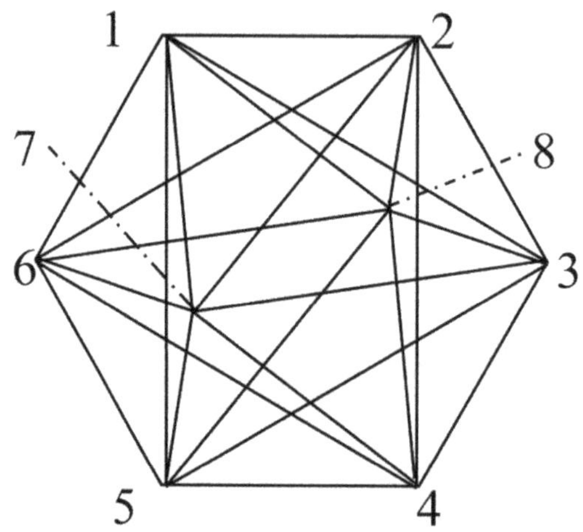

Figure 7. The symmetrical form of the 4-cross-polytope

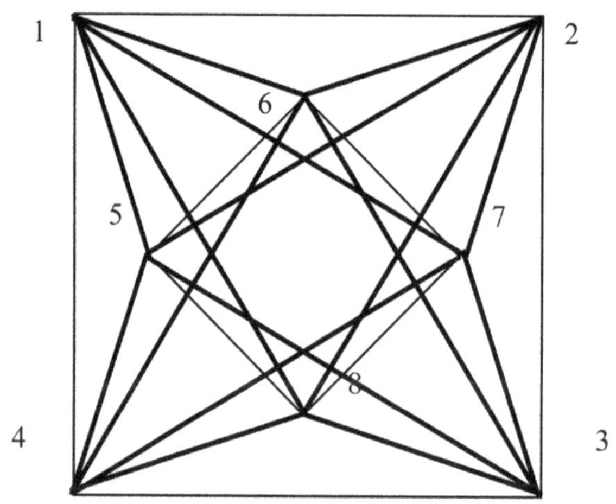

the vertices of the outer contour of the projection of the 4-cross- polytope of the projection of the edges perpendicular to the middle lines of projections. The points of their intersection give new vertices. Through these vertices, we draw straight lines parallel to the corresponding edges of the original projection. The intersection points of these lines give four more recent vertices of the enlarged projection of the 4-cross-polytope (Figure 8).

Obviously, the process of extending the 4-cross-polytope can be continued further. Easy to see that each time the size (length) of the edges of the next 4-cross-popytope from the size of the edges of the previous 4-cross-polytope

Figure 8. The hierarchical filling of the space by a 4-cross-polytope

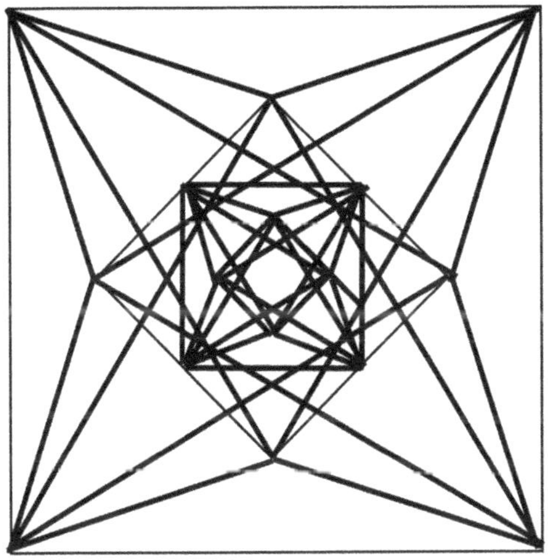

is increased by 3 times. Thus, the geometric progression coefficient of the hierarchical filling of the space by a 4-cross-polytope is 3.

In the same way, it is possible to construct a hierarchical filling of a space of dimension greater than four starting from some initial figure of this dimension. For example, let this figure be a 5-simplex. Its image on the plane is shown in Figure 9 (Zhizhin, 2014c).

This polytope has 6 of the vertices, 16 of the edges, 20 of the two-dimensional faces, 15 of the three-dimensional faces, 6 of the four-dimensional faces. Continuing the outer edges of the projection to the intersection with each

Figure 9. The polytope 5-simplex

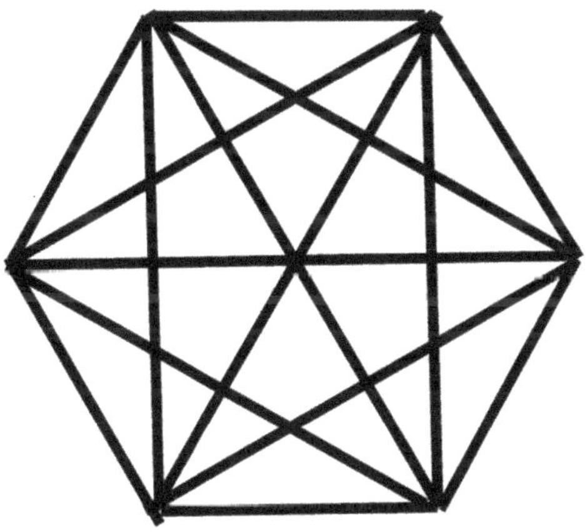

other, connecting the resulting vertices with edges, we obtain an enlarged projection of the 5-simplex (Figure 10).

Obviously, the process of extending the 5-simplex can be continued further. Easy to see that each time the size (length) of the edges of the next 5-simplex from the size of the edges of the previous 5-simplex is increased by $\sqrt{3}$ times. Thus, the geometric progression coefficient of the hierarchical filling of the space by a 4-simplex is $\sqrt{3}$.

In the second chapter, the adamantane molecule, which is the main part of the unit cell of the diamond, was considered. It was proved that the adamantane molecule has a dimensionality of 4 (Zhizhin, 2014d; Zhizhin, Khalaj & Diudea, 2016). The octahedron enters the adamantane molecule as a three-dimensional face. We can assume that the octahedron is the skeleton of the adamantane molecule (see Figure 15 in Chapter 2). Studying a set of cubic cells including adamantane molecules, one can establish the existence of scaling process in the diamond, i.e. formation of large-scale geometric configurations from the same figures of a smaller scale. Figure 11 shows the result of octahedron receiving on the basis of 8 cubes, each of them containing an octahedron 8 times smaller. It explains the existence of diamond crystals of macroscopic dimensions with the same form as a microscopic unit cell of the diamond.

Figure 10. The hierarchical filling of the space by a 5-simplex

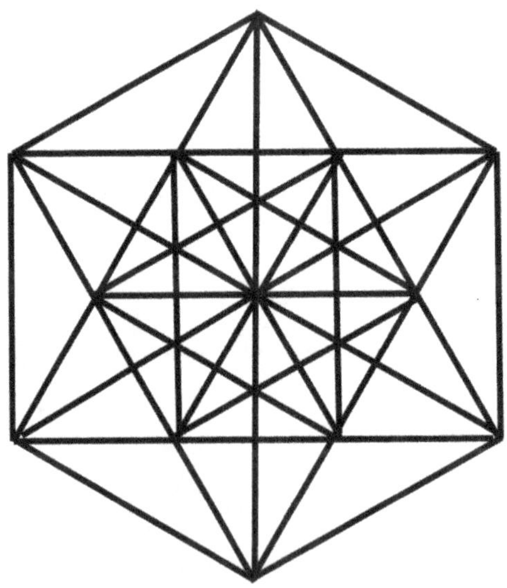

Figure 11. Scaling in a diamond

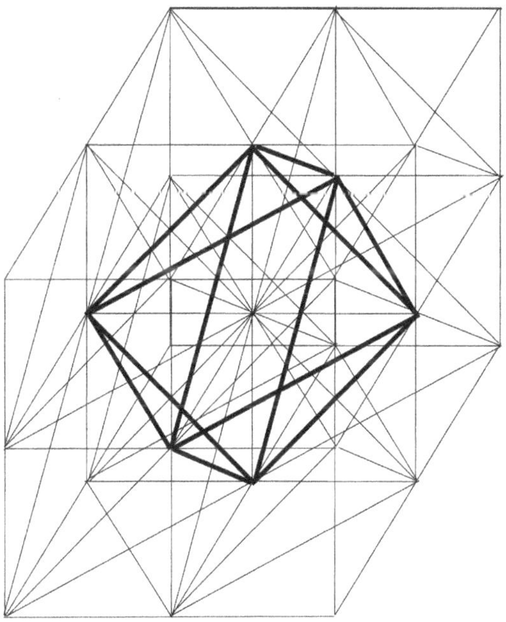

The increase of scale at that occurs in a discrete manner. The scale of diamond crystal increases $n^3(n = 1, 2, ...)$ times. The enlarged diamond crystal contains many carbon atoms, preserving the continuity of the medium and the shape of its microcrystalline sample.

POLYTOPIC PRISMAHEDRONS: STEREOHEDRONS OF THE *N*-DIMENSION SPACES

A characteristic feature of hierarchical filling of spaces is the removal of the vertices of a figure from each other during its expansion. Naturally, if the vertices of the figure are images of the atoms of the molecule, the removal of the atoms from each other should lead to their separation from each other. This will not happen if the expansion of the figures does not come from a single exceptional point, but in principle, from each point of space. Like the expansion of the Universe, each point of which is equal. Then next to any vertex are other vertices (atoms) and the destruction of matter does not occur. We see such a picture in the diffraction patterns of nanostructures (Shevchenko, Zhizhin & Mackay, 2013; Zhizhin & Diudea, 2016b). Since we can isolate points located at certain distances from each other, then at expansions of these points in space, periodic sets of vertices arise, as a consequence of the hierarchical filling of space from these points. This we also see in the diffraction patterns of nanostructures. The sets of vertices, as a result of a certain shift in any direction, as shown in Chapter 5, form a polytopic prismahedron. We show that the polytopic prismahedrons are the stereohedrons introduced by Delone to describe the partitions of n-dimensional spaces; polytopic prismahedrons allow filling the n-dimensional space without a gap between the prismahedrons (Zhizhin, 2015).

Theorem 1

Polytopic prizmahedrons dimension n perform the correct partition $\mathfrak{I}$ congruent space of dimension n, moreover, polytopic prizmahedrons included in the partition $\mathfrak{I}$ or have no common elements, or their common elements have the dimension m ($0 \leq m \leq n$ -1), but each of these polytopic prizmaedrons have elements of dimension n - 1, which do not belong to any of the neighboring politopic prizmahedrons.

Proof

Consider tetrahedral prism, i.e. a product tetrahedron by the segment (Chapter 5)

$$P_4^3(4F_3^2) \times P_2^1 = P_8^4[4P_6^3(3F_4^2, 2F_3^2), 2F_4^3(4F_3^2)]. \tag{1}$$

Here, the subscript in the polytope P and his face F indicates the number of vertices, and the superscript indicates the dimension of the corresponding polytope or faces. The right side of (1) describing the structural formula of the product, the facet indicated by the symbol of the polytope to specify which polytopes of dimension n - 1 is composed work polytopes. Thus, P_4^3 - the tetrahedron, P_2^1 - segment, P_6^3- triangular prism, P_3^2- triangle, P_4^2- quadrilateral, P_8^4 - tetrahedral prism. The dimension of the tetrahedral prism is equal to 4, it has 8 vertices, 16 edges, 14 faces two-dimensional, 6 three-dimensional faces (2 tetrahedrons, 4 triangular prisms). Image tetrahedral prism is shown in Figure 12.

We introduce one of the vertices of the tetrahedral prism origin of the four-dimensional space (x, y, z, t). Orient the coordinates, such as indicated

Figure 12. The tetrahedral prism

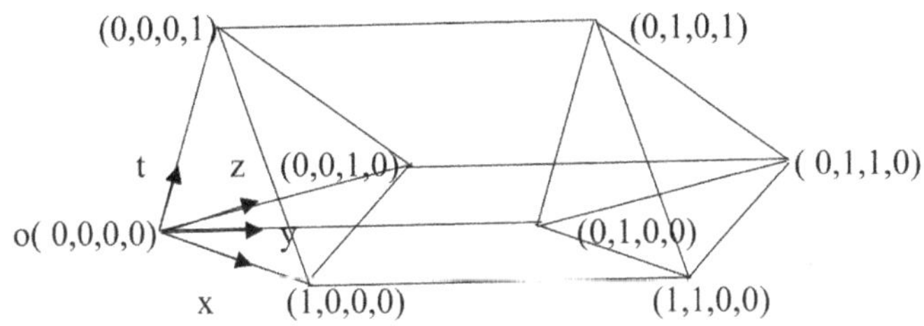

in Figure 12. Assume that the length of each edge is equal to 1. Then, each node tetrahedral prism can be associated with a set of integers (Figure 12). Translating tetrahedral prism along the coordinates x, y, z, t, we obtain the lattice vertices. Let A_0 tetrahedral prism with the values of vertex coordinates in Figure 12. Then

$$A_0 = [(0,0,0,0),(1,0,0,0),(0,1,0,0),(0,0,1,0),(0,0,0,1),(1,1,0,0),(0,1,1,0),(0,1,0,1)], \quad (2)$$

$$A_1 = A_0(x+1) = [(1,0,0,0),(2,0,0,0),(1,1,0,0),(1,0,1,0),(1,0,0,1),(2,1,0,0),(1,1,1,0),(1,1,0,1)],$$

$$A_2 = A_0(z+1) = [(0,0,1,0),(1,0,1,0),(0,1,1,0),(0,0,2,0),(0,0,1,1),(1,1,1,0),(0,1,2,0),(0,1,1,1)],$$

$$A_3 = A_0(x+1,z+1) = A_1(z+1) = [(1,0,1,0),(2,0,1,0),(1,1,1,0),(1,0,2,0),(1,0,1,1,),(2,1,1,0),(1,1,2,0),(1,1,1,1)],$$

$$A_4 = A_0(y+1,z+1) = A_2(y+1) = [(0,1,1,0),(1,1,1,0),(0,2,1,0),(0,1,2,0),(0,1,1,1),(1,2,1,0),(0,2,2,0),(0,2,1,1)],$$

$$A_5 = A_0(x+1,y+1,z+1) = A_3(y+1) = [(1,1,1,0),(2,1,1,0),(1,2.1.0),(1,1,2,0),(1,1,1,1),(2,2,1,0),(1,2,2,0),(1,2,1,1)],$$

$$A_6 = A_0(y+1) = [(0,1,0,0),(1,1,0,0),(0,2,0,0),(0,1,1,0),(0,1,0,1),(1,2,0,0),(0,2,1,0),(0,2,0,1)],$$

$$A_7 = A_0(y+1, \mathrm{x}+1) = A_1(y+1) =$$
$$[(1,1,0,0),(2,1,0,0),(1,2,0,0),(1,1,1,0),(1,1,0,1),(2,2,0,0),(1,2,1,0),(1,2,0,1)],$$

$$A_8 = A_0(\mathrm{t}+1) =$$
$$[(0,0,0,1),(1,0,0,1),(0,1,0,1),(0,0,1,1),(0,0,0,2),(1,1,0,1),(0,1,1,1),(0,1,0,2)],$$

$$A_9 = A_0(y+1, \mathrm{t}+1) = A_6(\mathrm{t}+1) =$$
$$[(0,1,0,1),(1,1,0,1),(0,2,0,1),(0,1,1,1),(0,1,0,2),(1,2,0,1),(0,2,1,1),(0,2,0,2)],$$

$$A_{10} = A_0(\mathrm{x}+1, y+1, \mathrm{t}+1) = A_7(\mathrm{t}+1) =$$
$$[(1,1,0,1),(2,1,0,1),(1,2,0,1),(1,1,1,1),(1,1,0,2),(2,2,0,1),(1,2,1,1),(1,2,0,2)],$$

$$A_{11} = A_0(\mathrm{x}+1, \mathrm{t}+1) = A_1(\mathrm{t}+1) =$$
$$[(1,0,0,1),(2,0,0,1),(1,1,0,1),(1,0,1,1),(1,0,0,2),(2,1,0,1),(1,1,1,1),(1,1,0,2)],$$

$$A_{12} = A_0(\mathrm{y}+1, \mathrm{z}+1) = A_2(\mathrm{t}+1) =$$
$$[(0,0,1,1),(1,0,1,1),(0,1,1,1),(0,0,2,1),(0,0,1,2),(1,1,1,1),(0,1,2,1),(0,1,1,2)],$$

$$A_{13} = A_0(\mathrm{z}+1, \mathrm{y}+1, t+1) = A_4(\mathrm{t}+1) =$$
$$[(0,1,1,1),(1,1,1,1),(0,2,1,1),(0,1,2,1),(0,2,2,1),(0,1,1,2),(1,2,1,1),(0,2,1,2)],$$

$$A_{14} = A_0(\mathrm{x}+1, \mathrm{y}+1, z+1, t+1) = A_5(\mathrm{t}+1) =$$
$$[(1,1,1,1),(2,1,1,1),(1,2,1,1),(1,1,2,1),(1,1,1,2),(2,2,1,1),(1,2,2,1),(1,2,1,2)],$$

$$A_{15} = A_0(\mathrm{x}+1, \mathrm{t}+1, z+1) = A_3(\mathrm{t}+1) =$$
$$[(1,0,1,1),(2,0,1,1),(1,1,1,1),(1,0,2,1),(1,0,1,2),(2,1,1,1),(1,1,2,1),(1,1,1,2)].$$

Representing a tetrahedral prisms $A_0 \div A_{15}$ dots in three-dimensional space, we get the hypercube. Moreover, the edges of the hypercube correspond to possible changes in the values of one of the coordinates of the vertices of the tetrahedral prism unit (Figure 13).

In addition, each edge of the hypercube in Figure 13 can be considered as an element of the overall two tetrahedral prisms, connected by an edge.

Figure 13. The hypercube from 16 tetrahedral prisms

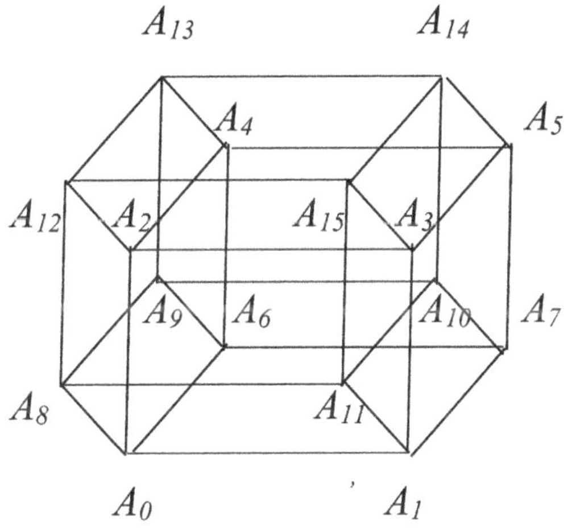

Using the coordinate expression tetrahedral prisms (2) can be analytically determined. In Table 1 geometry elements are common to each pair of tetrahedral prisms connected by an edge in Figure 13 are listed.

Diagonals flat faces in the hypercube correspond to a simultaneous change in the unit values of the two coordinates of the vertices of tetrahedral prisms (Table 2). Diagonals 8 cubes in the hypercube in Figure 13 correspond to a simultaneous change in the unit of some three coordinates. Common elements of tetrahedral prisms with such changes vertex coordinates is either a vertex or the empty set. When you change the same unit coordinates all four common elements in tetrahedral prisms not.

Between A_i type tetrahedral prisms arranged tetrahedral prism type B_i (see Figure 14) having a closest to them tetrahedral prisms A_i general quadrangular two-dimensional face.

Tetrahedrons prisms A_i and B_i have common edges, which are parties to the general quadrilateral faces. Tetrahedral prism A_i and B_i are connected symmetry transformation - turning the 180^0 around a common edge of the tetrahedron. Let B_0 tetrahedral prism type B_i, adjacent tetrahedral prism A_0, contains the x-axis, will translate the prism B_0 on edge length for all the coordinates x, y, z, t four-dimensional space. Then we obtain a lattice of tetrahedral prisms B_i.

$$B_0 = [(0,0,0,0),(0,1,0,0),(1,1,0,0),(1,0,0,0),(1,0,-1,0),\\ (1,0,0,-1),(1,1,0,-1),(1,1,-1,0)], \tag{3}$$

Figure 14. The partition space by tetrahedral prisms

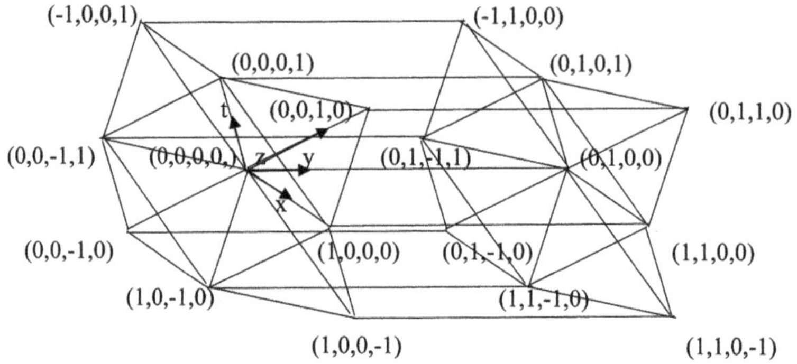

$B_1 = B_0(x+1) = [(1,0,0,0),(1,1,0,0),(2,1,0,0),(2,0,0,0),$
$(2,0,-1,0),(2,0,0,-1),(2,1,0,-1),(2,1,-1,0)],$

$B_2 = B_0(z+1) = [(0,0,1,0),(0,1,1,0),(1,1,1,0),$
$(1,0,1,0),(1,0,0,0),(1,0,1,-1),(1,1,1,-1),(1,1,0,0)],$

$B_3 = B_0(x+1,z+1) = B_1(z+1) =$
$[(1,0,1,0),(1,1,1,0),(2,1,1,0),(2,0,1,0),(2,0,0,0,),(2,0,1,-1),(2,1,1,-1),(2,1,0,0)],$

$B_4 = B_0(y+1,z+1) = B_2(y+1) =$
$[(0,1,1,0),(0,2,1,0),(1,2,1,0),(1,1,1,0),(1,1,0,0),(1,1,1,-1),(1,2,1,-1),(1,2,0,0)],$

$B_5 = B_0(\mathrm{x}+1,y+1,z+1) = B_3(y+1) =$
$[(1,1,1,0),(1,2,1,0),(2,2.1.0),(2,1,1,0),(2,1,0,0),(2,1,1,-1),(2,2,1,-1),(2,2,0,0)],$

$B_6 = B_0(y+1) = [(0,1,0,0),(0,2,0,0),(1,2,0,0),(1,1,0,0),$
$(1,1,-1,0),(1,1,0,-1),(1,2,0,-1),(1,2,-1,0)],$

$B_7 = B_0(y+1,\mathrm{x}+1) = B_1(y+1) =$
$[(1,1,0,0),(1,2,0,0),(2,2,0,0),(2,1,0,0),(2,1,-1,0),(2,1,0,-1),(2,2,0,-1),(2,2,-1,0)],$

$$B_8 = B_0(\mathrm{t}+1) =$$
$$[(0,0,0,1),(0,1,0,1),(1,1,0,1),(1,0,0,1),(1,0,-1,1),(1,0,0,1),(1,1,0,0),(1,1,-1,1)],$$

$$B_9 = B_0(y+1,\mathrm{t}+1) = B_6(\mathrm{t}+1) =$$
$$[(0,1,0,1),(0,2,0,1),(1,2,0,1),(1,1,0,1),(1,1,-1,1),(1,1,0,0),(1,2,0,0),(1,2,-1,1)],$$

$$B_{10} = B_0(\mathrm{x}+1,y+1,\mathrm{t}+1) = B_7(\mathrm{t}+1) =$$
$$[(1,1,0,1),(1,2,0,1),(2,2,0,1),(2,1,0,1),(2,1,-1,1),(2,1,0,0),(2,2,0,0),(2,2,-1,1)],$$

$$B_{11} = B_0(\mathrm{x}+1,\mathrm{t}+1) = B_1(\mathrm{t}+1) =$$
$$[(1,0,0,1),(1,1,0,1),(2,1,0,1),(2,0,0,1),(2,0,-1,1),(2,0,0,0),(2,1,0,0),(2,1,-1,1)],$$

$$B_{12} = B_0(\mathrm{y}+1,\mathrm{z}+1) = B_2(\mathrm{t}+1) =$$
$$[(0,0,1,1),(0,1,1,1),(1,1,1,1),(1,0,1,1),(1,0,0,1),(1,0,1,0),(1,1,1,0),(1,1,0,1)],$$

$$B_{13} = B_0(\mathrm{z}+1,\mathrm{y}+1,t+1) = B_4(\mathrm{t}+1) =$$
$$[(0,1,1,1),(0,2,1,1),(1,2,1,1),(1,1,1,1),(1,1,0,1),(1,1,1,0),(1,2,1,0),(1,2,0,1)],$$

$$B_{14} = B_0(\mathrm{x}+1,\mathrm{y}+1,z+1,t+1) = B_5(\mathrm{t}+1) =$$
$$[(1,1,1,1),(1,2,1,1),(2,2,1,1),(2,1,1,1),(2,1,0,1),(2,1,1,0),(2,2,1,0),(2,2,0,1)],$$

$$B_{15} = B_0(\mathrm{x}+1,\mathrm{t}+1,z+1) = B_3(\mathrm{t}+1) =$$
$$[(1,0,1,1),(1,1,1,1),(2,1,1,1),(2,0,1,1),(2,0,0,1),(2,0,1,0),(2,1,1,0),(2,1,0,1)].$$

It is presented tetrahedral prisms $B_0 \div B_{15}$ points in 4-dimensional space we get up hypercube in Figure 8 with change notation A_i on B_i. From the construction it is follows that tetrahedral prism B_i can have common elements such as tetrahedrons, edges and vertices. This is easily seen by analytical expressions (3). In addition, it follows from the construction (Figure 9) that the tetrahedral prism A_i and B_i can have common elements only tetrahedrons,

Table 1. Common elements of tetrahedral prisms

Edge of Hypercube	Common Element of Tetrahedral Prisms	Edge of Hypercube	Common Element of Tetrahedral Prisms
$A_0 A_1$ $A_0 A_2$ $A_0 A_6$ $A_0 A_8$ A_1A_3 A_7A_1 A_1A_{11} A_2A_3 A_2A_4 A_2A_{12} A_3A_5 A_3A_{15} A_4A_5 A_4A_6 A_4A_{13} A_7A_5	edge (1,0,0,0)(1,1,0,0) edge (0,1,1,0)(0,0,1,0) tetrahedron (0,1,0,0)(1,1,0,0)(0,1,0,1)(0,1,0,1) edge (0,1,0,1)(0,0,0.1) edge (1,0,1,0)(1,1,1,0) tetrahedron (1,1,0,0,)(2,1,0,0)(1,1,1,0)(1,1,0,1) edge (1,0,0,1)(1,1,0,1) edge (1,0,1,0)(1,1,0,0) tetrahedron (0,1,1,0)(1,1,1,0)(0,1,1,1)(0,1,2,0) edge (0,1,1,0)(0,0,1,1) tetrahedron (1,1,1,0)(2,1,1,0)(1,1,2,0)(0,1,1,1) edge (1,0,1,1)(1,1,1,1) edge (1,1,1,0)(1,2,1,0) edge (0,1,1,0)(0,2,1,0) vertex (0,1,1,1) edge (1,1,1,0)(1,2,1,0)	$A_5 A_{14}$ $A_6 A_7$ $A_6 A_9$ A_7A_{10} $A_8 A_{11}$ A_8A_9 A_8A_{12} A_9A_{10} A_9A_{13} $A_{11}A_{10}$ $A_{14}A_{10}$ $A_{11}A_{15}$ $A_{13}A_{12}$ $A_{12}A_{15}$ $A_{13}A_{14}$ $A_{14}A_{15}$	edge (1,1,1,1)(1,2,1,1) vertex (1,1,1,0) edge (0,1,0,1)(0,2,0,1) vertex (1,1,0,1) edge (1,0,0,1)(1,1,0,1) tetrahedron (0,1,0,1)(1,1,0,1)(0,1,1,1)(0,1,0,2) edge (0,0,1,1)(0,1,1,1) edge (1,1,0,1)(1,2,0,1) edge (1,2,1,1)(0,1,1,1) edge (1,2,0,1)(1,1,0,1) edge (1,1,1,1)(1,2,1,1) edge (1,0,1,1)(1,1,1,1) edge (0,1,1,1)(1,1,1,1) edge (1,0,1,1)(1,1,1,1) tetrahedron (1,1,1,1)(1,1,2,1)(1,2,1,1)(1,2,2,1) tetrahedron (1,1,1,1)(2,1,1,1)(1,1,2,1)(1,1,1,1)

edges and vertices. Triangular prisms which are present in tetrahedral prisms of both types cannot be common for these tetrahedral prisms.

They only deal with each other on a flat quadrilateral. This also easily seen from the formulas (2) and (3). This situation is significantly different from the conditions of normality, to accept in general theory stereohedrons. The provided construction proves that the tetrahedral prism is a fundamental area in the 4-dimensional space, as it completely fills the space in the translation of the prism in all the coordinates of space and turning it into 180^0. Since all the vertices of the tetrahedral prisms are equal and filling the space with their translation is uniform, then the partition created by them is the right one space. It is obvious that the nature of the mutual arrangement of polytopes will not change, if the three-dimensional polyhedron in a polytopic prism will attend any other convex polyhedron. Common elements polytopic prisms are themselves three-dimensional polyhedrons and elements of lower dimension. Three-dimensional prism here also are not common elements of polytopic prisms. Consider a complex polytopic prism, i.e., polytopic prismahedron resulting from the product of two polytopes. For example, let polytopic prismahedron is the product of a tetrahedron by a triangle. The result of this product is a polytope of dimension 5, which can be called tetrahedral prismahedron (Chapter 5).

Table 2. Common elements of tetrahedral prisms

Diagonal Flat Face of Hypercube	Common Element of Tetrahedral Prisms	Diagonal Flat Face of Hypercube	Common Element of Tetrahedral Prism
$A_8 A_{13}, A_8 A_{13}$	vertex (0,1,1,1)	$A_8 A_{10}, A_9 A_{11}$	vertex (1,1,1,1)
$A_6 A_2, A_0 A_4$	vertex (0,1,1,0)	$A_{14} A_7$	∅
$A_8 A_2$	edge (0,1,1,1)(0,0,1,1)	$A_5 A_{10}$	edge (1,1,1,1)(1,2,1,1)
$A_0 A_{12}$	∅	$A_{13}A_{14}, A_5 A_{15}$	vertex (1,1,1,1)
$A_4 A_9$	vertex (0,1,1,1)(0,0,1,1)	A_3A_{11}	vertex (1,0,1,1)(1,1,1,1)
A_6A_{13}	∅	$A_1 A_{15}$	∅
A_5A_{13}	edge (1,1,1,1)(1,2,1,1)	$A_7A_{11}, A_1 A_{10}$	vertex (1,1,0,1)
$A_4 A_{14}$	∅	$A_5 A_1, A_7 A_3$	vertex (1,1,1,0)
A_5A_{10}	edge (1,2,1,1)(1,1,1,1)	A_3A_{12}	edge (1,0,1,1)(1,1,1,1)
A_7A_{14}	∅	A_2A_{15}	∅
A_7A_9	edge (1,1,0,1)(1,2,0,1)	$A_{11}A_0$	∅
$A_6 A_{10}$	∅	A_1A_8	edge (1,0,0,1)(1,1,0,1)
A_9A_{14}	∅	$A_4A_3, A_5 A_2$	edge (1,1,1,0)
$A_{13}A_{10}$	edge (1,2,1,1)(1,1,1,1)	$A_1 A_6, A_0 A_7$	vertex (1,1,0,0)
A_5A_6	∅	$A_6 A_8, A_0 A_9$	vertex (0,1,0,1)
A_4A_7	edge (1,1,1,0)(1,2,1,0)	$A_{11}A_{14}, A_{10} A_{15}$	vertex (1,1,1,1)
$A_{11}A_{12}$	edge (1,1,1,1)(1,0,1,1)	A_8A_{15}	∅
$A_{14}A_{12}$	vertex (1,1,1,1)	$A_{13}A_{15}$	edge (1,1,2,1)(1,1,1,1)

$$P_4^3(4F_3^2)]\times P_3^2 = P_{12}^5\{4P_9^4[6F_6^3(3F_4^2,2F_3^2),2P_8^4[4F_6^3(3F_4^2,2F_3^2),2F_4^3(4F_3^2)]\}. \tag{4}$$

Polytopes (4) to be composed from 3 tetrahedral prisms P_8^4 and 4 triangular prismahedrons P_9^4. Its image it is shown in Figure 15.

This figure shows that the translation of the tetrahedral prismahedron on edge length in the coordinates 5-dimensional space to carry out to form between them of tetrahedral prismahedrons, rotated relative to the first in 180^0. In addition, each 4-dimensional tetrahedral prism belongs to only one tetrahedral prismahedron, i.e., in this case, four-dimensional tetrahedral prisms do not belong simultaneously two adjacent 5-dimensional tetrahedral prismahedrons. Obviously, the above regularities are performed when constructing spaces with the help of more complicated politopic prismahedrons.

This proves the theorem 1.

A NEW PARADIGM OF DISCRETE SYSTEMS

We recall briefly the main concepts and results of the classical theory of discrete systems. An example of the realization of constructing a space from congruent polytopes (18 Hilbert's problem) to be the crystalline structures

Figure 15. The tetrahedral prismahedron

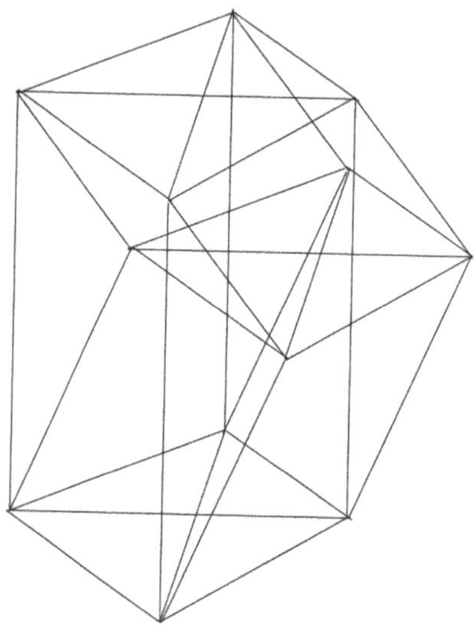

that are widespread in nature. A crystalline structure is called correct if the groups in all those space motions, like a rigid whole that combine this structure with itself, is discrete and has a fundamental domain. A group *G* of motions is said to be discrete, if there exists a point *A* and a positive number *r* such, that every point different from *A* and equivalent to *A* is relative to *G* (that is, the one into which the point *A* passes by the motion from *G*) lies no closer to *A* then on distance *r*. The fundamental domain of the group *G* is a set of points of space such that 1) all its points are not equivalent to each other with respect *G*, and 2) any point of the space is equivalent to an equivalent point of this domain with respect *G*.

Such groups *G* are called crystallographic. These classical definitions (Delone, Padurov & Alexandrov, 1934) implicitly assume that the considered spaces are infinite, although all crystal structures in nature have a boundary and a finite volume. It was first established by Fedorov (Fedorov, 1890) and somewhat later by Schönfles (Schönfles, 1891) that there are 17 on the two-dimensional plane, and 230 crystallographic groups in the three-dimensional space. If we add to the conditions of discreteness and finiteness of the fundamental domain the requirement that in the group *G* there exist an n-dimensional subgroup *T* of parallel transfers(if *G* is n-dimensional), then the proof of the finiteness of the number of such groups follows directly from the Frobenius argument (Frobenius, 1911). Evidence for the existence of such n-dimensional subgroups *T* was obtained in a number of papers

(Schönfles, 1891, Zasenkaus, 1948, Bieberbach, 1910, 1911, 1912). Delone (Delone,1937) introduced the concept about system of the distribution of discrete matter, based on two conditions: the existence of a finite smallest radius of the ball inside which there are no points of the system (the concept of an empty ball), and the existence of a radius of the ball inside which there is necessarily at least one point of the system. The question arises as to what are the convex polytopes that play the role of fundamental domains in n-dimensional Euclidean space. Based on the introduced notion of a discrete system Delone proved (Delone, 1961) that a normal partition of a space (polytopes are adjacent to each other along entire (n-1) -dimensional faces) for any n there are only a finite number of topologically different types of partition. In the definition of normal partitions is assumed that for every (n-1) face of polytope P in the normal decomposition there is one and only one other polytope S having this same face. If we take the group G and repeat a point A, then we obtain a regular system of points. The Dirichlet domains of its points form some regular Dirichlet decomposition connected with the group G. It is normal. Delone and Sandakova proved (Delone, Sandakova, 1961) that the stereohedrons of these partitions for a fixed n can only be of a finite number of topologically different types. The edge grids of the stereohedrons do not stretch to each other. If we do not require the normality of partitions into convex fundamental domains, then there can be infinitely many such topologically distinct partitions (Zamorzayev, 1965).

It is absolutely clear that the concept of discrete systems introduced by Delone clearly contradicts the existence of a scaling process, which has been proved experimentally. The concept of an empty ball does not allow a continuous reduction in scale. In this paper, we introduce an idea of a discrete system that agrees with scaling processes. We consider an arbitrary n-dimensional system of points in n-dimensional space. The system is discrete, i.e., between points there is always some distance, which can only tend asymptotically to zero, not reaching an exact equality to zero. The system has in the neighborhood of each point the hierarchical distribution of points (Mackay, 2001). The latter leads to the existence of a periodic distribution of points in space.

The introduction of the discrete system so introduced corresponds to the diffraction patterns of the nanostructures and to the scaling process discovered in recent years. It is essential that the lattice of points of this system allows for the separation in it, as shown in Chapter 4, of all Platonic solids, Bravais and Delone cells. Thus, this discrete system is universal. Based on this new conception of a discrete system, it was possible to obtain and systematize

(Chapter 5) polytopic prismahedrons of the higher dimension, which, as shown in this chapter, allow filling an n-dimensional space without gaps. The direct construction of polytopic prismahedrons showed that the obligatory condition accepted in the theory of the Delone stereohedrons (Tarasov, 1997), does not hold, the separation of vertices of neighboring stereohedrons along the n -1 plane (Theorem 1 of Chapter 7).

The discrete Delone system has so far not been able to obtain a single example of a stereohedron higher dimensional (Galiulin, 2003). Until no enumeration of the stereohedrons of Delone with dimension 3 of normal partitions has been obtained. There are only isolated examples of these stereohedrons (Shtogrin, 1973; Peter, 1981). Moreover, the use of discrete Delane systems can lead to incorrect results. For example, in the work of Ryzhkov, Shushbaev (Ryzhkov & Shushbaev, 1981) is asserted on the basis of ideas about these systems, that with the help of a 4-cross-polytope one can obtain the correct partition of the space 4*D*. However, a direct construction by the methods developed in this paper can show that this is not so.

The most complete variants of normal mono-, di-, and polyhedral partitions of the two-dimensional Euclidean plane are considered by Zhizhin (1993).

REFERENCES

Bieberbach, L. (1910). Uber die Bewegungsruppen des n-dimensionalen Euklidischen Räumes mit einem endlichen fundamentalbereich. *Gött. Nachr.*, 75 – 84.

Bieberbach, L. (1911). Uber die Bewegungsruppen der Euklidischen Räume. I. *Math. Ann.*, *70*(3), 207–336. doi:10.1007/BF01564500

Bieberbach, L. (1912). Uber die Bewegungsruppen des Euklidischen Räume. II. *Math. Ann.*, *72*, 400–412. doi:10.1007/BF01456724

Delone, B. (1937). Geometry of positive quadratic forms. *Uspekhi Mat. Nauk.*, *3*, 16–62.

Delone, B. (1961). Proof of the main theorem of the theory of stereohedrons. *Reports of the Academy of Sciences of the USSR*, *138*(6), 1270–1272.

Delone, B. (1969). To the eighteenth problem of Hilbert. In P. Aleksandrov (Ed.), *Problems of Hilbert* (pp. 200–203). Moscow: Science.

Delone, B., Padurov, N., & Aleksandrov, A. (1934). *Mathematical foundations of the structural analysis of crystals*. Leningrad: ONTI.

Delone, B., & Sandakova, N. (1961). Theory of stereohedrons. *Proceedings of the Mathematical Institute. V.A. Steklov, 64*, 28 – 51.

Eiji, A., Yanfa, Y., & Pennycook, S. J. (2004). Quasicrystals as cluster aggregates. *Nature Materials, 3*(11), 759–767. doi:10.1038/nmat1244 PMID:15516956

Fedorov, E. S. (1885). *The beginning of the doctrine of figures*. Saint Petersburg: Academy of Sciences.

Fischer, M. E., & Pfeuty, P. (1972). Critical behavior of the anisotropic n-vector model. *Physical Review B: Condensed Matter and Materials Physics, 6*(5), 1889–1891. doi:10.1103/PhysRevB.6.1889

Frobenius, G. (1911). Ueber die unzerlegbaren diskreten Bewegungsgruppen. *Sitzb. Preuss. Akad. Wiss., Phys.- Math. K, 1*, 654–655.

Galiulin, R. (2003). Systems B.N. Delone as the mathematical foundation of a discrete world. *J. Computed. Mat. and Math. Physics, 43*(6), 790–801.

Hilbert D. (1901). *Gesammelte Abhandlungen*. Academic Press.

Kadanoff, L. P. (1966). Scaling Laws For Ising Models Near $\tau_c^* \tau_c^*$. *Physics, 2*, 263–272.

Krapivsky, P. L., Redner, S., & Ben-Naim, E. (2010). *A Kinetic View of Statistical Physics*. Cambridge, UK: Cambridge University Press. doi:10.1017/CBO9780511780516

Mackay, A. (2001). On Complexity. *Crystallography Reports, 46*(4), 524–526. doi:10.1134/1.1387117

Mukhopadhyay, N.K. (1993). Diffraction studies of icosahedral phases in $Al_{70}Fe_{20}W_{10}$. *Journal of Non-Crystalline Solids, 153-154*, 1193 – 1197.

Peter, E. (1986). *Geometric Crystallography*. Dordrecht: Springer.

Ryzhkov, S. S., & Shusbaev, S. Sh. (1981). The structur of the L-partition for the second perfect lattice. *Mat. Collection, 146*(2), 218–231.

Schönfless. (1891). *Kristallsysteme und Kristallstruktur*. Liepzig.

Shevchenko, V., Zhizhin, G., & Mackay, A. (2013). On the structure of the quasicrystals in the high dimension space. In M. V. Diudea (Ed.), *Diamond and Related Nanostructures*. Dordrecht: Springer. doi:10.1007/978-94-007-6371-5_17

Stogrin, M. I. (1973). Correct Dirichlet-Voronoi decompositions for the second triclinic group. *Trudy Mat. Institute of Steklov*, *123*, 128.

Tarasov A.S. (1997). Complexity of convex stereohedrons. *Mathematic Notes, 61*(5), 797 – 800.

Wilson, R. G. (1971). Renormmalization group and critical phenomena. I. Renormalization group and the Kadanoff scaling picture. *Physical Review B: Condensed Matter and Materials Physics*, *4*, 3174–3183. doi:10.1103/PhysRevB.4.3174

Zamorzayev, A.M. (n.d.). On abnormal regular partitions of Euclidean space. *Reports of the Academy of Sciences of the USSR, 161*(1), 30-31.

Zang, X., & Kelton, K. (1993). High-order crystal approximant alloys $Ti_{54}Zr_{26}Ni_{20}$. *Journal of Non-Crystalline Solids, 153-154*, 114–118. doi:10.1016/0022-3093(93)90325-R

Zhizhin, G. V. (1993). *Parquetages of equal triangles, adjoining on whole sides*. St. Petersburg: Polytechnic.

Zhizhin, G. V. (2010). *Geometrical bases of the dissipative structures*. St. Petersburg: Polytechnic.

Zhizhin, G. V. (2012, October). *Hierarchical filling of spaces with polytopes.* Paper presented at "St. Petersburg Scientific Forum: Science and Human Progress". 7th St.- Petersburg meeting of Nobel Prize laureates, St. Petersburg, Russia.

Zhizhin, G. V. (2014a). The fractal nature of disproportionate phases. *10th all-Russian Scientific School "Mathematical research in the natural sciences"*. Apatity: Geological institute KSC RAS.

Zhizhin, G. V. (2014b). Incommensurable and fluctuating structures in the terrestrial space. *Biosphere*, *3*, 211–221.

Zhizhin, G. V. (2014c). *World – 4D*. St. Petersburg: Polytechnic Service.

Zhizhin, G. V. (2014d). On the higher dimension in nature. *Biosphere*, *6*(4), 313–318.

Zhizhin, G. V. (2015, November). *Polytopic prismahedrons – fundamental regions of the n-dimension nanostructures.* Paper Presented at the International conference "Nanoscience in Chemistry, Physics, Biology and Mathematics", Cluj-Napoca, Romania.

Zhizhin, G. V., & Diudea, M. V. (2016). Space of Nanoworld. In M. V. Putz & M. C. Mirica (Eds.), *Sustainable Nanosystems, Development, Properties, and Applications* (pp. 214–236). New York: IGI Global.

Zhizhin, G. V., Khalaj, Z., & Diudea, M. V. (2016). Geometrical and topology dimensions of the diamond. In A. R. Ashrafi & M. V. Diudea (Eds.), *Distance, symmetry and topology in carbon nanomaterials* (pp. 167–188). New York: Springer. doi:10.1007/978-3-319-31584-3_12

KEY TERMS AND DEFINITIONS

New Paradigm of Discrete n-Dimension World: The elementary cells of the translational filling of the n-dimensional space are the polytopic prismahedrons, the stereohedrons of which Delone spoke, but he did not give a single concrete example of stereohedron. Polytopic prismahedrons, filling the n-dimensional space, as shown by direct construction, can have common elements in the entire range of dimensions up to the dimension of the facets, or do not have any common elements. In addition to the translational filling of the n-dimensional space, there is a hierarchical filling of the space, which is inextricably linked with the scaling process, i.e. discrete scale change of the figure. Translational filling of space can be combined with hierarchical filling of space. In this case, in principle, in each point of space there is an asymptotic decrease or increase in the scale of the figure (as an expansion of the Universe from each of its points). Delone's provisions on an "empty" ball, the finite minimum and maximum distances between the points of a discrete system, are not used. Thus, Hilbert's problem acquires a completely new content.

Conclusion

The study showed that almost all elements of the periodic system of Mendeleev to announce in the chemical compounds the valence is higher than the valence determined by the number of the group in the periodic system. To explain this in the scientific literature resorts to assumptions about various mechanisms of electron interaction. Practical all these compounds form molecules of higher dimension, determined by the Euler-Poincaré equation. The existence of closed objects (molecules) of higher dimension in a space of lower dimension does not contradict Riemann's geometry, which assumes the boundedness of a space with a given dimension. Images of many molecules of higher dimension are given. In particular, new images of biomolecules having high dimensional values were obtained.

Conformations of chiral molecules of higher dimension (glucose, tartaric acid) are considered. The constructed images of higher dimension made it possible to explain the experimentally observed differences in the properties of these conformations.

To obtain materials with new properties, the question of filling the space with polytopes that are images of complex molecules is important. If polytopes are congruent, we arrive at the well-known 18th Hilbert problem. In the case of a multidimensional space, Delone introduced representations of stereohedron that could fill a multidimensional space without gaps as a result of the translation of the stereohedron along the directions of the space coordinates. However, until now no stereohedron has been introduced that could solve this problem even for a four-dimensional space. In this paper, the question of the filling of a multidimensional space by the translation of polytopes of higher dimension was solved by obtaining a product of polytopes. It is the products of polytopes have the properties necessary for filling multidimensional space with translation, i.e. they are stereohedrons. The products of polytopes are not well-studied simplicial polytopes. Therefore, their study is of independent interest. In the work various products of polytopes, called polytopic prismahedrons, are systematized, their analytical expressions are obtained, and their specific images are given. The direct construction of space by means of translation of polytopic prismahedrons, found that the condition accepted in the theory of the Delone stereohedrons for the contact of stereohedrons only over facets is not satisfied. The polytopic prismahedrons, when they fill the space, can contact each other with respect to elements of various dimensions from 0 to the dimension of the facets.

In the description of discrete systems, there is no need to introduce the concept of an empty ball, which is necessary in the Delone theory. The concept of an empty sphere of finite diameter, in which there are no vertices (atoms) of polytopes, contradicts, for example, the process of scaling, i.e., continuous change of scale in condensed systems, discovered in recent decades. This contradiction is clearly seen when observing diffraction patterns of nanostructures. In these diffraction patterns you can find out how filling the space with polytopes of higher dimension during their translation, and hierarchical filling of space, accompanied by a change in the scale of the figures, i.e. scaling of figures. In this connection, the paper has studied the different way of the hierarchical filling of spaces with regular and semi - regular polytopes of higher dimension.

One of the important characteristics of polytopes is the construction of dual polytopes. Therefore, continuing the study of polytopic prismahedrons, the paper analyzes the polytopes dual to polytopic prismahedrons. An absolutely new class of polytopes – poly-incidence polytopes is found in which simultaneously there are edges with different degrees of incidence to the three-dimensional faces of polytopes.

Proving the higher dimensionality of many molecules, the question naturally arose of the possibility of solving the Schrödinger equation in a space of higher dimension, since known electron orbitals of atoms were obtained as a result of solving the Schrödinger equation in a three-dimensional metric space. It is found that the Schrodinger equation has a solution in a space of higher dimension. The mistake of the previous researchers is that they unreasonably complicated the law of the Coulomb interaction of electric charges. This led them to a paradoxical conclusion about the impossibility of the atomic structure of matter in spaces of higher dimension, because there was no solution to the Schrödinger equation under these conditions. A solution of the Schrödinger equation in a four-dimensional metric space was obtained. From this solution, in particular, follows the possibility of the existence of more quantum cells in different orbitals of the atoms. The latter can explain the greater chemical activity of atoms than the Mendeleyev table.

In general, we can say that this work creates the basis for a new stereochemistry - stereochemistry in spaces of higher dimension.

Appendix

SOLUTION OF THE SCHRÖDINGER EQUATION IN A *P*-DIMENSIONAL METRIC SPACE

The Stationary Schrödinger Equation in a *p*-Dimensional Metric Space

The stationary Schrödinger equation in a *p*-dimensional metric space has view

$$-\frac{\hbar}{2m}\nabla_p^2\Psi + U\Psi = E\Psi. \tag{1}$$

where Ψ. - the wave function, U - the potential energy of an electron (the function of radius *r*), *E* – the kinetic energy of electron, ∇_p^2 - Laplacian in *p*-dimensional metric space, *m* – the mass of electron, $\hbar$.- Planck constant.

The potential energy *U(r)* – this potential energy of the Coulomb attraction of an electron to the nucleus and can be written in explicit form

$$U(r) = -\frac{Ze^2}{r}. \tag{2}$$

where eZ - ion`s charge, *e* – electron`s charge.

It is essential that the form of the potential energy (2) does not depend on the dimensionality of space and it is defined as a function inversely proportional to the radius in the first degree. Since in a space of any dimension, the potential energy is a function of the distance between the charges (of the electron and the nucleus). The incorrectness in previous studies of the solution of the Schrödinger equation in spaces of higher dimension is that the authors of these works (Büchel, 1963; Freeman, 1969; Gurevich & Mostepanenko,

1971) recorded potential energy without a physical basis as a function of the radius to the power of p -2. Such an entry goes to formula (2) only in three-dimensional space, but in a space of the higher dimension this record is not true and leads to an absurd conclusion about the impossibility of the existence of a discrete atomic structure in a space of higher dimension.

Equation (1), having transferred to the atomic system of units, taking into account (2), in the Cartesian coordinate system has the form

$$\frac{\partial^2\Psi}{\partial x_1^2}+\frac{\partial^2\Psi}{\partial x_2^2}+\ldots+\frac{\partial^2\Psi}{\partial x_p^2}+\left(\frac{c}{r}+2E\right)\Psi=0. \qquad (3)$$

where c – constant, $r=\sqrt{x_1^2+\ldots+x_p^2}$.

For further analysis, it is convenient to go over into a spherical coordinate system, where the radius of one of the p variables. The remaining p - 1 variables are the angles between the vector of length r to the point of location of the particle and the unit vectors of the Cartesian system of coordinates.

It can be noted that Gurevich & Mostepanenko (1971) were confused with coordinate systems. For example, the velocity vector in the n-dimensional space has n + 1 components in their work. Babenko (2015), considering n-dimensional space, describes the one-dimensional Schrödinger equation, forgetting the remaining variables.

In an n-dimensional space, the analytic connection between the coordinates is expressed by equalities (Hobson, 1931)

$$x_1=r\sin\alpha_1\sin\alpha_2\ldots\sin\alpha_{p-1},$$

$$x_2=r\cos\alpha_1\sin\alpha_2\ldots\sin\alpha_{p-1},$$

$$x_3=r\cos\alpha_2\sin\alpha_3\ldots\sin\alpha_{p-1}, \qquad (4)$$

$$x_{p-1}=r\cos\alpha_{p-2}\sin\alpha_{p-2},$$

$$x_p=r\cos\alpha_{p-1}.$$

The Laplacian in the Schrodinger equation in a spherical coordinate system in p-dimensional space can be written in the form

$$\nabla_p^2 = \frac{1}{L_1 L_2 L_3 \ldots L_p}\left[\frac{\partial}{\partial r}\left(\frac{L_2 L_3 \ldots L_p}{L_1}\frac{\partial \Psi}{\partial r}\right) + \frac{\partial}{\partial \alpha_1}\left(\frac{L_1 L_3 \ldots L_p}{L_2}\frac{\partial \Psi}{\partial \alpha_1}\right) + \frac{\partial}{\partial \alpha_2}\left(\frac{L_1 L_2 L_4 \ldots L_p}{L_3}\frac{\partial \Psi}{\partial \alpha_2}\right) + \right.$$

$$\left. \ldots + \frac{\partial}{\partial \alpha_{p-1}}\left(\frac{L_1 L_2 \ldots L_{p-1}}{L_p}\frac{\partial \Psi}{\partial \alpha_{p-1}}\right)\right]. \quad (5)$$

where L_i - the Lame coefficients, $i = 1, 2, \ldots, p$,

$$L_1 = \sqrt{\left(\frac{\partial x_1}{\partial r}\right)^2 + \left(\frac{\partial x_2}{\partial r}\right)^2 + \ldots + \left(\frac{x_p}{\partial r}\right)^2},$$

$$L_2 = \sqrt{\left(\frac{\partial x_1}{\partial \alpha_1}\right)^2 + \left(\frac{\partial x_2}{\partial \alpha_1}\right)^2 + \ldots + \left(\frac{x_p}{\partial \alpha_1}\right)^2}, \quad (6)$$

$$L_p = \sqrt{\left(\frac{\partial x_1}{\partial \alpha_{p-1}}\right)^2 + \left(\frac{\partial x_2}{\partial \alpha_{p-1}}\right)^2 + \ldots + \left(\frac{x_p}{\partial \alpha_{p-1}}\right)^2}.$$

Differentiating the equality (4) with respect to the independent variables, substituting the derivatives in the Lame coefficients (6), we obtain after the transformations

$$\begin{aligned} &L_1 = 1, L_2 = r \sin \alpha_2 \ldots \sin \alpha_{p-1}, L_3 = r \sin \alpha_3 \ldots \sin \alpha_{p-1}, \\ &L_4 - r \sin \alpha_4 \ldots \sin \alpha_{p-1}, \ldots, L_p = 1. \end{aligned} \quad (7)$$

The Schrödinger equation in a spherical coordinate system in p-dimensional space has view

$$\nabla_p^2\Psi + \left(\frac{2}{r} + 2E\right)\Psi = 0. \tag{8}$$

Where $\nabla_p^2\Psi$ is expressed using equations (5) – (7).

A solution of equation (8) can be obtained by separating the variables assuming that

$$\Psi\left(r,\alpha_1,...,\alpha_{p-1}\right) = R\left(r\right)\Phi_1\left(\alpha_1\right)\Phi_2\left(\alpha_2\right)...\Phi_{p-1}\left(\alpha_{p-1}\right). \tag{9}$$

We confine ourselves to the case of four-dimensional space, $p = 4$. Than Schrödinger equation (8) has view

$$\frac{1}{r^3}\frac{\partial}{\partial r}\left(r^3\frac{\partial\Psi}{\partial r}\right) + \frac{1}{r^2\sin\alpha_2{}^2\sin\alpha_3{}^2}\frac{\partial^2\Psi}{\partial\alpha_1{}^2} + \frac{1}{r^2\sin\alpha_2\sin\alpha_3{}^2}\left(\sin\alpha_2\frac{\partial\Psi}{\partial\alpha_2}\right) + \frac{1}{r^2\sin\alpha_3{}^2}\frac{\partial}{\partial\alpha_3}\left(\sin\alpha_3{}^2\frac{\partial\Psi}{\partial\alpha_3}\right) + 2\frac{\Psi}{r} + 2E\Psi = 0. \tag{10}$$

The Decision of the Stationary Schrödinger Equation in a *p*-Dimension Metric Space

We seek the wave function in the form

$$\Psi\left(r,\alpha_1,\alpha_2,\alpha_3\right) = R\left(r\right)\Phi_1\left(\alpha_1\right)\Phi_2\left(\alpha_2\right)\Phi_3\left(\alpha_3\right). \tag{11}$$

We substitute (11) into (10)

$$\frac{1}{rR(r)}\frac{d}{dr}\left(r^3\frac{dR(r)}{dr}\right) + 2r^2\left(\frac{1}{r} + E\right) = -\frac{1}{\Phi_1(\alpha_1)\sin\alpha_2{}^2\sin\alpha_3{}^2}\frac{d^2\Phi_1(\alpha_1)}{d\alpha_1{}^2} - \frac{1}{\Phi_2(\alpha_2)\sin\alpha_2\sin\alpha_3{}^2}\frac{d}{d\alpha_2}\left(\sin\alpha_2\frac{d\Phi_2(\alpha_2)}{d\alpha_2}\right) - \frac{1}{\Phi_3(\alpha_3)\sin\alpha_3{}^2}\frac{d}{d\alpha_3}\left(\sin\alpha_3{}^2\frac{d\Phi_3(\alpha_3)}{d\alpha_3}\right). \tag{12}$$

The left part of (12) is depended at radius *r*, and right part is dependent only at angles.

Therefore, both parts equality (12) equal to same constant C_1. From left part of (12) we have

$$\frac{1}{rR(r)}\frac{d}{dr}\left(r^3\frac{dR(r)}{dr}\right)+2r^2\left(\frac{1}{r}+E\right)=C_1. \tag{13}$$

From right part of (12) we have

$$C_1=-\frac{1}{\Phi_1(\alpha_1)\sin\alpha_2^{\,2}\sin\alpha_3^{\,2}}\frac{d^2\Phi_1(\alpha_1)}{d\alpha_1^{\,2}}-\frac{1}{\Phi_2(\alpha_2)\sin\alpha_2\sin\alpha_3^{\,2}}\frac{d}{d\alpha_2}\left(\sin\alpha_2\frac{d\Phi_2(\alpha_2)}{d\alpha_2}\right)-\frac{1}{\Phi_3(\alpha_3)\sin\alpha_3^{\,2}}\frac{d}{d\alpha_3}\left(\sin\alpha_3^{\,2}\frac{d\Phi_3(\alpha_3)}{d\alpha_3}\right). \tag{14}$$

The equation (14) we rewrite as follows

$$C_1\sin\alpha_3^{\,2}+\frac{1}{\Phi_3(\alpha_3)}\frac{d}{d\alpha_3}\left(\sin\alpha_3^{\,2}\frac{d\Phi_3(\alpha_3)}{d\alpha_3}\right)=-\frac{1}{\Phi_1(\alpha_1)\sin\alpha_2^{\,2}}\frac{d^2\Phi_1(\alpha_1)}{d\alpha_1^{\,2}}-\frac{1}{\Phi_2(\alpha_2)\sin\alpha_2}\frac{d}{d\alpha_2}\left(\sin\alpha_2\frac{d\Phi_2(\alpha_2)}{d\alpha_2}\right). \tag{15}$$

The left part of (15) is depended at angles only, and the right part of (15) is depended at angles only. Therefore, both parts of (15) equal to same constant C_2. From the left part (15) we have

$$C_1\sin\alpha_3^{\,2}+\frac{1}{\Phi_3(\alpha_3)}\frac{d}{d\alpha_3}\left(\sin\alpha_3^{\,2}\frac{d\Phi_3(\alpha_3)}{d\alpha_3}\right)=C_2. \tag{16}$$

From the right part of (15) we have

$$-\frac{1}{\Phi_1(\alpha_1)}\frac{d^2\Phi_1(\alpha_1)}{d\alpha_1^{\,2}}=C_2\sin\alpha_2^{\,2}+\frac{\sin\alpha_2}{\Phi_2(\alpha_2)}\frac{d}{d\alpha_2}\left(\sin\alpha_2\frac{d\Phi_2(\alpha_2)}{d\alpha_2}\right). \tag{17}$$

The left part of (17) is depended only at α_1, and right part of (17) is depended only at α_2.Therefore, both parts equal to same constant C_3. From the left part of (17) we have

$$-\frac{1}{\Phi_1(\alpha_1)}\frac{d^2\Phi_1(\alpha_1)}{d\alpha_1^{\ 2}}=C_3. \tag{18}$$

From the right part of (17) we have

$$C_2\sin\alpha_2^{\ 2}+\frac{\sin\alpha_2}{\Phi_2(\alpha_2)}\frac{d}{d\alpha_2}\left(\sin\alpha_2\frac{d\Phi_2(\alpha_2)}{d\alpha_2}\right)=\mathrm{C}_3\,. \tag{19}$$

The equation (18) have decision

$$\Phi_1(\alpha_1)=Ae^{\pm i\sqrt{C_3}\alpha_1}. \tag{20}$$

Since $\Phi_1(0)=\Phi_1(2\pi)$, then $A=Ae^{\pm i\sqrt{C_3}2\pi}$, i.e.

$$e^{\pm i\sqrt{C_3}2\pi}=1. \tag{21}$$

Consequently, $\cos(2\pi\sqrt{C_3})\pm i\sin(2\pi\sqrt{C_3})=1.$
It may be only if $\sqrt{C_3}=0,\pm1,\pm2,\ldots.$
So we got the first quantum number

$$m_1=\sqrt{C_3}=0,\pm1,\pm2,\ldots. \tag{22}$$

The constant A is defined from condition of normalization

$$\int_0^{2\pi}\Phi_1^*(\alpha_1)\Phi_1(\alpha_1)d\alpha_1=1=A^2\int_0^{2\pi}e^{-i\sqrt{C_3}\alpha_1}e^{i\sqrt{C_3}\alpha_1}d\alpha_1=A^2 2\pi.$$

Consequently, $A=\frac{1}{\sqrt{2\pi}}$. Thus, the function it is known

$$\Phi_1(\alpha_1)=\frac{1}{\sqrt{2\pi}}e^{-im_1\alpha_1},m_1=\sqrt{C_3}. \tag{23}$$

We write the equation (19) in the form

$$\Phi_2(\alpha_2)[C_2 - \frac{m_1^2}{\sin\alpha_2{}^2}] + \frac{1}{\sin\alpha_2}\frac{d}{d\alpha_2}\left(\sin\alpha_2 \frac{d\Phi_2(\alpha_2)}{d\alpha_2}\right) = 0. \tag{24}$$

Equation (24) is the adjoined Legendre equation. In the case $C_2 = l(l+1), l$ are integers, its solutions are the adjoined Legendre polynomials (Landau & Lifshitz, 1963)

$$P_l^{m_1}(\cos\alpha_2) = \frac{1}{2^l l!}\sin\alpha_2{}^{m_1}\frac{d^{m_1+l}}{(d\cos\alpha_2)^{m_1+l}}(\cos\alpha_2{}^2 - 1)^l. \tag{25}$$

So we got second quantum number *l*. Equation (16) can be rewritten in the form

$$\Phi_3(\alpha_3)\left(C_1 - l\frac{l+1}{\sin\alpha_3{}^2}\right) + \frac{1}{\sin\alpha_3{}^2}\frac{d}{d\alpha_3}\left(\sin\alpha_3{}^2 \frac{d\Phi_3(\alpha_3)}{d\alpha_3}\right) = 0. \tag{26}$$

Since $\sin\alpha_3{}^2$ is a positive function varying from 0 to 1, then we introduce the average value of this function τ in the range α_3 from 0 to 2π

$$\tau = \frac{1}{2\pi}\int_0^{2\pi}\sin\alpha_3{}^2 d\alpha_3 = \frac{1}{2}.$$

Then the equation (26) have the form

$$C_1 - 2l(l+1) = -\frac{1}{\Phi_3(\alpha_3)}\frac{d^2\Phi_3(\alpha_3)}{d\alpha_3{}^2}. \tag{27}$$

The decision of this equation is

$$\Phi_3(\alpha_3) = Be^{\pm i\alpha_3\sqrt{C_1 - 2l(l+1)}2\pi}. \tag{28}$$

Since $\Phi_3(0) = \Phi_3(2\pi)$, so from (28) we have

$$B = Be^{\pm i\sqrt{C_1 - 2l(l+1)}2\pi}, \text{i.e. } e^{\pm i\sqrt{C_1 - 2l(l+1)}2\pi} = 1.$$

Consequently, $\cos(2\pi\sqrt{C_1 - 2l(l+1)}) \pm i\sin(2\pi\sqrt{C_1 - 2l(l+1)}) = 1.$
It may be only if $\sqrt{C_1 - 2l(l+1)} = 0, \pm1, \pm2, \ldots.$
So we got the second quantum number

$$m_2 = \sqrt{C_1 - 2l(l+1)} = 0, \pm1, \pm2, \ldots. \quad (29)$$

The constant *B* is defined from condition of normalization (see equation (20))

$$\Phi_3(\alpha_3) = \frac{1}{\sqrt{2\pi}} e^{-im_2\alpha_3}. \quad (30)$$

Now consider the equation for the radius (13). We perform differentiation in it

$$r^2 \frac{d^2R}{dr^2} + 3r\frac{dR}{dr} + R(2r + 2r^2E - 2l(l+1) - m_2^2) = 0. \quad (31)$$

We seek the solution of the equation (31) in the form of a series in powers of *r*

$$\mu = (-E)^{1/2}, E < 0, b_j \text{ are numerical coefficients.} \quad (32)$$

Substituting (32) into (31), performing differentiation and combining summands with the same powers *r*, we obtain

$$\sum_j b_j r^{l+j}\left((l+j)^2 + 2(l+j) - 2l(l+1) - m_2^2\right) = \sum_j b_j r^{l+j+1}\left(\mu(2l+2j+3) - 2\right). \quad (33)$$

From equation (33) follows the recurrence formula for the coefficients

$$b_{j+1} = \frac{\mu(2l + 2j + 3) - 2}{(l + j + 1)^2 + 2(l + j + 1) - 2l(l + 1) - m_2^2} b_j. \tag{34}$$

In view of the need to obtain a limited physical meaningful solution, the values of the coefficients, starting from a certain value of *j* must vanish. This is possible if the numerator on the right-hand side of (34) is zero

$$\mu(2l + 2j + 3) - 2 = 0. \tag{35}$$

From equation (35) follows that

$$\mu = \sqrt{-E} = \frac{2}{2l + 2j + 3}. \tag{36}$$

We introduce the notation $n = l + j + 1$. From (36) equation we have

$$E = -\frac{2}{(2n + 1)^2}. \tag{37}$$

For big *n* formula (37) go to the formula of Bore at energy of electron.

The radial function is determined by the equality (32) of setting different values *j*. For example, with *j* = 0

$$R(r) = \mathrm{e}^{-\mu r} r^l b_0, b_0 = \frac{1}{\int_0^\infty e^{-\mu r} r^l dr}.$$

with *j* = *1*

$$R(r) = e^{-\mu r} r^l b_0 \left(1 + r \frac{\mu(2l + 3) - 2}{(l + 1)^2 + 2(l + 1) - 2l(l + 1) - m_2^2} \right),$$

$$b_0 = \frac{1}{\int_0^\infty e^{-\mu r} r^l \left(1 + r \frac{\mu(2l + 3) - 2}{(l + 1)^2 + 2(l + 1) - 2l(l + 1) - \mathrm{m}_2^2} \right) dr}.$$

Thus, one can obtain various approximations for the function $R(r)$..

Consideration of Decision of the Stationary Schrödinger Equation in a *p*-Dimensional Metric Space

Thus, it is proved that the Schrodinger equation has a solution in a space of higher dimension. Consequently, the discrete nature of matter exists in a space of higher dimension. For example, in four-dimensional space the atom has besides a spin quantum number yet of four more quantum numbers. The main quantum number n, which determines the average distance of the electron from the nucleus. Orbital quantum number l, characterizing the momentum of an electron. The first magnetic quantum number m_1, characterizing the possible value of the projection of the angular momentum of the electron on the axis z in a magnetic field. The second magnetic quantum number m_2 characterizing the possible value of the projection of the momentum of the electron on the fourth axis (t) in a magnetic field. The main quantum number can take the values 1, 2, 3, ... The orbital quantum number can take the values 0, 1, 2, ...

Magnetic quantum numbers can take the values 0, ± 1, ± 2, ± 3,...

The values of the orbital quantum number l correspond to form of electron orbital. Taking into account the large dimensionality of the neighborhood of the nucleus in the atom, we should expect an increase in the number of quantum cells in orbitals *p, d, f* while maintaining their shape. For example, not 6 but 8 electrons (for each quantum cell with a corresponding pair of electrons per axis) can be on the orbital *p*. In Chapters 1 and 2, when analyzing chemical elements and molecules formed by them, it was shown that many elements exhibit a valence in chemical reactions much higher than the valence determined by their place in the periodic table (group number). To explain this, we resort to assumptions about various possible mechanisms of electron interaction: atomic and molecular hybridization, repulsion of divided and unpaired electron pairs, and the involvement of electrons inside the electron shells in the chemical interactions. In addition, physical experiments at high pressures reveal a sharp change in the structure of compounds that are also not explainable from the point of view of standard valence.

On the other hand, the analysis of real forms of molecules shows that their dimensionality, determined by the shape of the corresponding convex body, whose vertices are atoms, is often higher than three.

Undoubtedly, the possible loss of the number of quantum cells in orbitals *p, d, f* should not lead to a change in the number of electrons in the atomic structure in neutral atoms. The number of electrons under these conditions is of course equal to the charge of the nucleus. But the possible increase in the number of vacant quantum cells contributes to an increase in the chemical activity of atoms, allows the creation of complex chemical compounds. This we see, for example, in biology. A detailed analysis of the molecules of complex compounds from the point of view of solutions of Schrödinger equations in a space of higher dimension still awaits its researchers.

REFERENCES

Babenko, Y. I. (2015). Antispace. In I.N. Toganov & Y.I. Babenko (Eds.), Anti-time and antispace. St. Petersburg: TIN.

Büchel, W. (1963). Warum hat ger Raume drei Dimensionen? *Physilalische Blatter*, *19*(12), 547–549. doi:10.1002/phbl.19630191204

Freeman, I. M. (1969). Why is Space Three-Dimensional? *American Journal of Physics*, *37*, 1222–1224. doi:10.1119/1.1975283

Gurevich, L., & Mostepanenko, V. (1971). On the Existence of Atoms in *n*-Dimensional Space. *Physics Letters. [Part A]*, *35*(3), 201–202. doi:10.1016/0375-9601(71)90148-4

Hobson, E. W. (1931). *The theory of spherical and allipsoidal harmonics*. Cambridge, UK: University Press.

Landau, L. D., & Lifshitz, E. M. (1963). Quantum mechanics. Nonrelativistic theory. Moscow: Gos. Ed. Physico-Mathematical Lit.

Related Readings

To continue IGI Global's long-standing tradition of advancing innovation through emerging research, please find below a compiled list of recommended IGI Global book chapters and journal articles in the areas of oxide-based catalysts, nanosheets, and wet chemical method. These related readings will provide additional information and guidance to further enrich your knowledge and assist you with your own research.

Abu Bakar, W. A., Abdullah, W. N., Ali, R., & Mokhtar, W. N. (2016). Polymolybdate Supported Nano Catalyst for Desulfurization of Diesel. In T. Saleh (Ed.), *Applying Nanotechnology to the Desulfurization Process in Petroleum Engineering* (pp. 263–280). Hershey, PA: IGI Global. doi:10.4018/978-1-4666-9545-0.ch009

Ahmad, W. (2016). Sulfur in Petroleum: Petroleum Desulfurization Techniques. In T. Saleh (Ed.), *Applying Nanotechnology to the Desulfurization Process in Petroleum Engineering* (pp. 1–52). Hershey, PA: IGI Global. doi:10.4018/978-1-4666-9545-0.ch001

Aïssa, B., & Khayyat, M. M. (2014). Self-Healing Materials Systems as a Way for Damage Mitigation in Composites Structures Caused by Orbital Space Debris. In M. Bououdina & J. Davim (Eds.), *Handbook of Research on Nanoscience, Nanotechnology, and Advanced Materials* (pp. 1–25). Hershey, PA: IGI Global. doi:10.4018/978-1-4666-5824-0.ch001

Akbari, E., Buntat, Z., Ahmadi, M. T., Karimi, H., & Khaledian, M. (2017). GAS Sensor Modelling and Simulation. In M. Ahmadi, R. Ismail, & S. Anwar (Eds.), *Handbook of Research on Nanoelectronic Sensor Modeling and Applications* (pp. 70–116). Hershey, PA: IGI Global. doi:10.4018/978-1-5225-0736-9.ch004

Akbari, E., Enzevaee, A., Karimi, H., Ahmadi, M. T., & Buntat, Z. (2017). Graphene-Based Gas Sensor Theoretical Framework. In M. Ahmadi, R. Ismail, & S. Anwar (Eds.), *Handbook of Research on Nanoelectronic Sensor Modeling and Applications* (pp. 117–149). Hershey, PA: IGI Global. doi:10.4018/978-1-5225-0736-9.ch005

Al-Najar, B. T., & Bououdina, M. (2016). Bioinspired Nanoparticles for Efficient Drug Delivery System. In M. Bououdina (Ed.), *Emerging Research on Bioinspired Materials Engineering* (pp. 69–103). Hershey, PA: IGI Global. doi:10.4018/978-1-4666-9811-6.ch003

AlMegren, H. A., Gonzalez-Cortes, S., Huang, Y., Chen, H., Qian, Y., Alkinany, M., & Xiao, T. et al. (2016). Preparation of Deep Hydrodesulfurzation Catalysts for Diesel Fuel using Organic Matrix Decomposition Method. In H. Al-Megren & T. Xiao (Eds.), *Petrochemical Catalyst Materials, Processes, and Emerging Technologies* (pp. 216–253). Hershey, PA: IGI Global. doi:10.4018/978-1-4666-9975-5.ch009

Alshammari, A., Kalevaru, V. N., Bagabas, A., & Martin, A. (2016). Production of Ethylene and its Commercial Importance in the Global Market. In H. Al-Megren & T. Xiao (Eds.), *Petrochemical Catalyst Materials, Processes, and Emerging Technologies* (pp. 82–115). Hershey, PA: IGI Global. doi:10.4018/978-1-4666-9975-5.ch004

Anwar, S. (2017). Wireless Nanosensor Networks: Prospects and Challenges. In M. Ahmadi, R. Ismail, & S. Anwar (Eds.), *Handbook of Research on Nanoelectronic Sensor Modeling and Applications* (pp. 505–511). Hershey, PA: IGI Global. doi:10.4018/978-1-5225-0736-9.ch017

Arafat, M. T., & Li, X. (2016). Functional Coatings for Bone Tissue Engineering. In A. Zuzuarregui & M. Morant-Miñana (Eds.), *Research Perspectives on Functional Micro- and Nanoscale Coatings* (pp. 240–264). Hershey, PA: IGI Global. doi:10.4018/978-1-5225-0066-7.ch009

Arokiyaraj, S., Saravanan, M., Bharanidharan, R., Islam, V. I., Bououdina, M., & Vincent, S. (2016). Green Synthesis of Metallic Nanoparticles Using Plant Compounds and Their Applications: Metallic Nanoparticles Synthesis Using Plants. In M. Bououdina (Ed.), *Emerging Research on Bioinspired Materials Engineering* (pp. 1–34). Hershey, PA: IGI Global. doi:10.4018/978-1-4666-9811-6.ch001

Balachandran, P. V., & Rondinelli, J. M. (2016). Informatics-Based Approaches for Accelerated Discovery of Functional Materials. In S. Datta & J. Davim (Eds.), *Computational Approaches to Materials Design: Theoretical and Practical Aspects* (pp. 192–223). Hershey, PA: IGI Global. doi:10.4018/978-1-5225-0290-6.ch007

Bamufleh, H. S., Noureldin, M. M., & El-Halwagi, M. M. (2016). Sustainable Process Integration in the Petrochemical Industries. In H. Al-Megren & T. Xiao (Eds.), *Petrochemical Catalyst Materials, Processes, and Emerging Technologies* (pp. 150–163). Hershey, PA: IGI Global. doi:10.4018/978-1-4666-9975-5.ch006

Banerjee, S., Gautam, R. K., Gautam, P. K., Jaiswal, A., & Chattopadhyaya, M. C. (2016). Recent Trends and Advancement in Nanotechnology for Water and Wastewater Treatment: Nanotechnological Approach for Water Purification. In A. Khitab & W. Anwar (Eds.), *Advanced Research on Nanotechnology for Civil Engineering Applications* (pp. 208–252). Hershey, PA: IGI Global. doi:10.4018/978-1-5225-0344-6.ch007

Barbero, C. A., & Yslas, E. I. (2017). Ecotoxicity Effects of Nanomaterials on Aquatic Organisms: Nanotoxicology of Materials on Aquatic Organisms. In S. Joo (Ed.), *Applying Nanotechnology for Environmental Sustainability* (pp. 330–351). Hershey, PA: IGI Global. doi:10.4018/978-1-5225-0585-3.ch014

Barbhuiya, S. (2014). Applications of Nanomaterials in Construction Industry. In M. Bououdina & J. Davim (Eds.), *Handbook of Research on Nanoscience, Nanotechnology, and Advanced Materials* (pp. 164–175). Hershey, PA: IGI Global. doi:10.4018/978-1-4666-5824-0.ch008

Barbu, M. C., Reh, R., & Çavdar, A. D. (2014). Non-Wood Lignocellulosic Composites. In A. Aguilera & J. Davim (Eds.), *Research Developments in Wood Engineering and Technology* (pp. 281–319). Hershey, PA: IGI Global. doi:10.4018/978-1-4666-4554-7.ch008

Barbu, M. C., Reh, R., & Irle, M. (2014). Wood-Based Composites. In A. Aguilera & J. Davim (Eds.), *Research Developments in Wood Engineering and Technology* (pp. 1–45). Hershey, PA: IGI Global. doi:10.4018/978-1-4666-4554-7.ch001

Bashir, R., & Chisti, H. (2015). Nanotechnology for Environmental Control and Remediation. In M. Shah, M. Bhat, & J. Davim (Eds.), *Nanotechnology Applications for Improvements in Energy Efficiency and Environmental Management* (pp. 156–183). Hershey, PA: IGI Global. doi:10.4018/978-1-4666-6304-6.ch006

Bayir, E., Bilgi, E., & Urkmez, A. S. (2014). Implementation of Nanoparticles in Cancer Therapy. In M. Bououdina & J. Davim (Eds.), *Handbook of Research on Nanoscience, Nanotechnology, and Advanced Materials* (pp. 447–491). Hershey, PA: IGI Global. doi:10.4018/978-1-4666-5824-0.ch018

Benjamin, S., Unni, K. N., Priji, P., & Wright, A. G. (2017). Biogenesis of Conjugated Linoleic Acids. In S. Benjamin (Ed.), *Examining the Development, Regulation, and Consumption of Functional Foods* (pp. 1–28). Hershey, PA: IGI Global. doi:10.4018/978-1-5225-0607-2.ch001

Berger, L. (2015). Tribology of Thermally Sprayed Coatings in the Al2O3-Cr2O3-TiO2 System. In M. Roy & J. Davim (Eds.), *Thermal Sprayed Coatings and their Tribological Performances* (pp. 227–267). Hershey, PA: IGI Global. doi:10.4018/978-1-4666-7489-9.ch008

Bhat, M. A., Nayak, B. K., Nanda, A., & Lone, I. H. (2015). Nanotechnology, Metal Nanoparticles, and Biomedical Applications of Nanotechnology. In M. Shah, M. Bhat, & J. Davim (Eds.), *Nanotechnology Applications for Improvements in Energy Efficiency and Environmental Management* (pp. 116–155). Hershey, PA: IGI Global. doi:10.4018/978-1-4666-6304-6.ch005

Bhutto, A. W., Abro, R., Abbas, T., Yu, G., & Chen, X. (2016). Desulphurization of Fuel Oils Using Ionic Liquids. In H. Al-Megren & T. Xiao (Eds.), *Petrochemical Catalyst Materials, Processes, and Emerging Technologies* (pp. 254–284). Hershey, PA: IGI Global. doi:10.4018/978-1-4666-9975-5.ch010

Bodratti, A. M., He, Z., Tsianou, M., Cheng, C., & Alexandridis, P. (2015). Product Design Applied to Formulated Products: A Course on Their Design and Development that Integrates Knowledge of Materials Chemistry, (Nano)Structure and Functional Properties. *International Journal of Quality Assurance in Engineering and Technology Education*, *4*(3), 21–43. doi:10.4018/IJQAETE.2015070102

Bogataj, D., & Drobne, D. (2017). Control of Perishable Goods in Cold Logistic Chains by Bionanosensors. In S. Joo (Ed.), *Applying Nanotechnology for Environmental Sustainability* (pp. 376–402). Hershey, PA: IGI Global. doi:10.4018/978-1-5225-0585-3.ch016

Bolboacă, S. D., & Jäntschi, L. (2017). Characteristic Polynomial in Assessment of Carbon-Nano Structures. In M. Putz & M. Mirica (Eds.), *Sustainable Nanosystems Development, Properties, and Applications* (pp. 122–147). Hershey, PA: IGI Global. doi:10.4018/978-1-5225-0492-4.ch004

Bouloudenine, M., & Bououdina, M. (2016). Toxic Effects of Engineered Nanoparticles on Living Cells. In M. Bououdina (Ed.), *Emerging Research on Bioinspired Materials Engineering* (pp. 35–68). Hershey, PA: IGI Global. doi:10.4018/978-1-4666-9811-6.ch002

Boumaza, N., Benouaz, T., & Goumri-Said, S. (2014). Understanding the Numerical Resolution of Perturbed Soliton Propagation in Single Mode Optical Fiber. In M. Bououdina & J. Davim (Eds.), *Handbook of Research on Nanoscience, Nanotechnology, and Advanced Materials* (pp. 492–504). Hershey, PA: IGI Global. doi:10.4018/978-1-4666-5824-0.ch019

Brimmo, A., & Emziane, M. (2014). Carbon Nanotubes for Photovoltaics. In M. Bououdina & J. Davim (Eds.), *Handbook of Research on Nanoscience, Nanotechnology, and Advanced Materials* (pp. 268–311). Hershey, PA: IGI Global. doi:10.4018/978-1-4666-5824-0.ch012

Brunetti, A., Sellaro, M., Drioli, E., & Barbieri, G. (2016). Membrane Engineering and its Role in Oil Refining and Petrochemical Industry. In H. Al-Megren & T. Xiao (Eds.), *Petrochemical Catalyst Materials, Processes, and Emerging Technologies* (pp. 116–149). Hershey, PA: IGI Global. doi:10.4018/978-1-4666-9975-5.ch005

Choi, H., Son, M., Bae, J., & Choi, H. (2017). Nanotechnology in Engineered Membranes: Innovative Membrane Material for Water-Energy Nexus. In S. Joo (Ed.), *Applying Nanotechnology for Environmental Sustainability* (pp. 50–71). Hershey, PA: IGI Global. doi:10.4018/978-1-5225-0585-3.ch003

del Valle-Zermeño, R., Chimenos, J. M., & Formosa, J. (2016). Flue Gas Desulfurization: Processes and Technologies. In T. Saleh (Ed.), *Applying Nanotechnology to the Desulfurization Process in Petroleum Engineering* (pp. 337–377). Hershey, PA: IGI Global. doi:10.4018/978-1-4666-9545-0.ch011

Deng, Y., & Liu, S. (2016). Catalysis with Room Temperature Ionic Liquids Mediated Metal Nanoparticles. In H. Al-Megren & T. Xiao (Eds.), *Petrochemical Catalyst Materials, Processes, and Emerging Technologies* (pp. 285–329). Hershey, PA: IGI Global. doi:10.4018/978-1-4666-9975-5.ch011

Dimitratos, N., Villa, A., Chan-Thaw, C. E., Hammond, C., & Prati, L. (2016). Valorisation of Glycerol to Fine Chemicals and Fuels. In H. Al-Megren & T. Xiao (Eds.), *Petrochemical Catalyst Materials, Processes, and Emerging Technologies* (pp. 352–384). Hershey, PA: IGI Global. doi:10.4018/978-1-4666-9975-5.ch013

Dixit, A. K., & Awasthi, R. (2015). EDM Process Parameters Optimization for Al-TiO2 Nano Composite. *International Journal of Materials Forming and Machining Processes*, *2*(2), 17–30. doi:10.4018/IJMFMP.2015070102

Elrawemi, M., Blunt, L., Fleming, L., Sweeney, F., Robbins, D., & Bird, D. (2015). Metrology of Al2O3 Barrier Film for Flexible CIGS Solar Cells. *International Journal of Energy Optimization and Engineering*, *4*(4), 46–60. doi:10.4018/IJEOE.2015100104

Emam, A. N., Mansour, A. S., Girgis, E., & Mohamed, M. B. (2017). Hybrid Nanostructures: Synthesis and Physicochemical Characterizations of Plasmonic Nanocomposites. In S. Joo (Ed.), *Applying Nanotechnology for Environmental Sustainability* (pp. 231–275). Hershey, PA: IGI Global. doi:10.4018/978-1-5225-0585-3.ch011

Emam, A. N., Mansour, A. S., Girgis, E., & Mohamed, M. B. (2017). Hybrid Plasmonic Nanostructures: Environmental Impact and Applications. In S. Joo (Ed.), *Applying Nanotechnology for Environmental Sustainability* (pp. 276–293). Hershey, PA: IGI Global. doi:10.4018/978-1-5225-0585-3.ch012

Emziane, M., & Yoosuf, R. (2014). In2X3 (X=S, Se, Te) Semiconductor Thin Films: Fabrication, Properties, and Applications. In M. Bououdina & J. Davim (Eds.), *Handbook of Research on Nanoscience, Nanotechnology, and Advanced Materials* (pp. 226–267). Hershey, PA: IGI Global. doi:10.4018/978-1-4666-5824-0.ch011

Fathi, A., Azizian, S., & Sharifan, N. (2017). Sensors and Amplifiers: Sensor Output Signal Amplification Systems. In M. Ahmadi, R. Ismail, & S. Anwar (Eds.), *Handbook of Research on Nanoelectronic Sensor Modeling and Applications* (pp. 423–504). Hershey, PA: IGI Global. doi:10.4018/978-1-5225-0736-9.ch016

Fathi, A., & Hassanzadazar, M. (2017). CNT as a Sensor Platform. In M. Ahmadi, R. Ismail, & S. Anwar (Eds.), *Handbook of Research on Nanoelectronic Sensor Modeling and Applications* (pp. 1–18). Hershey, PA: IGI Global. doi:10.4018/978-1-5225-0736-9.ch001

Ferbinteanu, M., Ramanantoanina, H., & Cimpoesu, F. (2017). Case Studies in the Challenge of Properties Design at Nanoscale: Bonding Mechanisms and causal Relationship. In M. Putz & M. Mirica (Eds.), *Sustainable Nanosystems Development, Properties, and Applications* (pp. 148–184). Hershey, PA: IGI Global. doi:10.4018/978-1-5225-0492-4.ch005

Fernández, S. M., & Marijuan, A. G. (2016). High-Tech Applications of Functional Coatings: Functional Coatings and Photovoltaic. In A. Zuzuarregui & M. Morant-Miñana (Eds.), *Research Perspectives on Functional Micro- and Nanoscale Coatings* (pp. 289–317). Hershey, PA: IGI Global. doi:10.4018/978-1-5225-0066-7.ch011

Fjodorova, N., Novic, M., Diankova, T., & Ostanen, A. (2016). The Nano-Sized TiO2 Dispersions for Mass Coloration of Polyimide Fibers: The Nano-Sized TiO2 for Mass Coloration. *Journal of Nanotoxicology and Nanomedicine*, *1*(1), 29–44. doi:10.4018/JNN.2016010103

Florea, L., Diamond, D., & Benito-Lopez, F. (2016). Opto-Smart Systems in Microfluidics. In A. Zuzuarregui & M. Morant-Miñana (Eds.), *Research Perspectives on Functional Micro- and Nanoscale Coatings* (pp. 265–288). Hershey, PA: IGI Global. doi:10.4018/978-1-5225-0066-7.ch010

Gaines, T. W., Williams, K. R., & Wagener, K. B. (2016). ADMET: Functionalized Polyolefins. In H. Al-Megren & T. Xiao (Eds.), *Petrochemical Catalyst Materials, Processes, and Emerging Technologies* (pp. 1–21). Hershey, PA: IGI Global. doi:10.4018/978-1-4666-9975-5.ch001

Galian, R. E., & Pérez-Prieto, J. (2016). Synergism at the Nanoscale: Photoactive Semiconductor Nanoparticles and their Organic Ligands. In A. Zuzuarregui & M. Morant-Miñana (Eds.), *Research Perspectives on Functional Micro- and Nanoscale Coatings* (pp. 42–77). Hershey, PA: IGI Global. doi:10.4018/978-1-5225-0066-7.ch003

Gaspar-Cunha, A., & Covas, J. A. (2014). An Engineering Scale-Up Approach using Multi-Objective Optimization. *International Journal of Natural Computing Research*, *4*(1), 17–30. doi:10.4018/ijncr.2014010102

Gaurina-Medjimurec, N., & Pasic, B. (2014). Risk Due to Pipe Sticking. In D. Matanovic, N. Gaurina-Medjimurec, & K. Simon (Eds.), *Risk Analysis for Prevention of Hazardous Situations in Petroleum and Natural Gas Engineering* (pp. 47–72). Hershey, PA: IGI Global. doi:10.4018/978-1-4666-4777-0.ch003

Gaurina-Medjimurec, N., & Pasic, B. (2014). Risk Due to Wellbore Instability. In D. Matanovic, N. Gaurina-Medjimurec, & K. Simon (Eds.), *Risk Analysis for Prevention of Hazardous Situations in Petroleum and Natural Gas Engineering* (pp. 23–46). Hershey, PA: IGI Global. doi:10.4018/978-1-4666-4777-0.ch002

Ge, H., Tang, M., & Wen, X. (2016). Ni/ZnO Nano Sorbent for Reactive Adsorption Desulfurization of Refinery Oil Streams. In T. Saleh (Ed.), *Applying Nanotechnology to the Desulfurization Process in Petroleum Engineering* (pp. 216–239). Hershey, PA: IGI Global. doi:10.4018/978-1-4666-9545-0.ch007

Golden, T. D., Tientong, J., & Mohamed, A. M. (2016). Electrodeposition of Nickel-Molybdenum (Ni-Mo) Alloys for Corrosion Protection in Harsh Environments. In A. Zuzuarregui & M. Morant-Miñana (Eds.), *Research Perspectives on Functional Micro- and Nanoscale Coatings* (pp. 369–395). Hershey, PA: IGI Global. doi:10.4018/978-1-5225-0066-7.ch014

Gopal, S., & Al-Hazmi, M. H. (2016). Advances in Catalytic Technologies for Selective Oxidation of Lower Alkanes. In H. Al-Megren & T. Xiao (Eds.), *Petrochemical Catalyst Materials, Processes, and Emerging Technologies* (pp. 22–52). Hershey, PA: IGI Global. doi:10.4018/978-1-4666-9975-5.ch002

Gupta, A. K., Dey, A., & Mukhopadhyay, A. K. (2016). Micromechanical and Finite Element Modeling for Composites. In S. Datta & J. Davim (Eds.), *Computational Approaches to Materials Design: Theoretical and Practical Aspects* (pp. 101–162). Hershey, PA: IGI Global. doi:10.4018/978-1-5225-0290-6.ch005

Han, B., Liu, W., & Zhao, D. (2017). In-Situ Oxidative Degradation of Emerging Contaminants in Soil and Groundwater Using a New Class of Stabilized MnO2 Nanoparticles. In S. Joo (Ed.), *Applying Nanotechnology for Environmental Sustainability* (pp. 112–136). Hershey, PA: IGI Global. doi:10.4018/978-1-5225-0585-3.ch006

Holberg, S. (2016). Non-Hydrolyzed Resins for Organic-Inorganic Hybrid Coatings: Functional Coating Films by Moisture Curing. In A. Zuzuarregui & M. Morant-Miñana (Eds.), *Research Perspectives on Functional Micro- and Nanoscale Coatings* (pp. 105–135). Hershey, PA: IGI Global. doi:10.4018/978-1-5225-0066-7.ch005

Hrncevic, L. (2014). Petroleum Industry Environmental Performance and Risk. In D. Matanovic, N. Gaurina-Medjimurec, & K. Simon (Eds.), *Risk Analysis for Prevention of Hazardous Situations in Petroleum and Natural Gas Engineering* (pp. 358–387). Hershey, PA: IGI Global. doi:10.4018/978-1-4666-4777-0.ch016

Hu, Y., Peng, Y., Liu, W., Zhao, D., & Fu, J. (2017). Removal of Emerging Contaminants from Water and Wastewater Using Nanofiltration Technology. In S. Joo (Ed.), *Applying Nanotechnology for Environmental Sustainability* (pp. 72–91). Hershey, PA: IGI Global. doi:10.4018/978-1-5225-0585-3.ch004

Huirache-Acuña, R., Alonso-Nuñez, G., Rivera-Muñoz, E. M., Gutierrez, O., & Pawelec, B. (2016). Trimetallic Sulfide Catalysts for Hydrodesulfurization. In T. Saleh (Ed.), *Applying Nanotechnology to the Desulfurization Process in Petroleum Engineering* (pp. 240–262). Hershey, PA: IGI Global. doi:10.4018/978-1-4666-9545-0.ch008

Huyen, P. T. (2017). Clay Minerals Converted to Porous Materials and Their Application: Challenge and Perspective. In T. Kobayashi (Ed.), *Applied Environmental Materials Science for Sustainability* (pp. 141–164). Hershey, PA: IGI Global. doi:10.4018/978-1-5225-1971-3.ch007

Jayavarthanan, R., Nanda, A., & Bhat, M. A. (2017). The Impact of Nanotechnology on Environment. In B. Nayak, A. Nanda, & M. Bhat (Eds.), *Integrating Biologically-Inspired Nanotechnology into Medical Practice* (pp. 153–193). Hershey, PA: IGI Global. doi:10.4018/978-1-5225-0610-2.ch007

Julião, D., Ribeiro, S., de Castro, B., Cunha-Silva, L., & Balula, S. S. (2016). Polyoxometalates-Based Nanocatalysts for Production of Sulfur-Free Diesel. In T. Saleh (Ed.), *Applying Nanotechnology to the Desulfurization Process in Petroleum Engineering* (pp. 426–458). Hershey, PA: IGI Global. doi:10.4018/978-1-4666-9545-0.ch014

Kamaja, C. K., Rajaperumal, M., Boukherroub, R., & Shelke, M. V. (2014). Silicon Nanostructures-Graphene Nanocomposites: Efficient Materials for Energy Conversion and Storage. In M. Bououdina & J. Davim (Eds.), *Handbook of Research on Nanoscience, Nanotechnology, and Advanced Materials* (pp. 176–195). Hershey, PA: IGI Global. doi:10.4018/978-1-4666-5824-0.ch009

Kannoorpatti, K., & Surovtseva, D. (2015). Integrating Industry Research in Pedagogical Practice: A Case of Teaching Microbial Corrosion in Wet Tropics. In H. Lim (Ed.), *Handbook of Research on Recent Developments in Materials Science and Corrosion Engineering Education* (pp. 254–272). Hershey, PA: IGI Global. doi:10.4018/978-1-4666-8183-5.ch013

Kanoun, M. B., & Goumri-Said, S. (2014). Theoretical Assessment of the Mechanical, Electronic, and Vibrational Properties of the Paramagnetic Insulating Cerium Dioxide and Investigation of Intrinsic Defects. In M. Bououdina & J. Davim (Eds.), *Handbook of Research on Nanoscience, Nanotechnology, and Advanced Materials* (pp. 431–446). Hershey, PA: IGI Global. doi:10.4018/978-1-4666-5824-0.ch017

Karimi, H., Rahmani, R., Akbari, E., Darabi, A. C., Rahmani, M., Ahmadi, M. T., & Anbari, S. (2017). Modeling of Sensing Layer of Surface Acoustic-Wave-Based Gas Sensors. In M. Ahmadi, R. Ismail, & S. Anwar (Eds.), *Handbook of Research on Nanoelectronic Sensor Modeling and Applications* (pp. 224–243). Hershey, PA: IGI Global. doi:10.4018/978-1-5225-0736-9.ch009

Karimi, H., Rahmani, R., Akbari, E., Rahmani, M., & Ahamdi, M. T. (2017). Optimization of Current-Voltage Characteristics of Graphene-Based Biosensors. In M. Ahmadi, R. Ismail, & S. Anwar (Eds.), *Handbook of Research on Nanoelectronic Sensor Modeling and Applications* (pp. 244–264). Hershey, PA: IGI Global. doi:10.4018/978-1-5225-0736-9.ch010

Kasani, H., Ahmadi, M. T., Khoda-Bakhsh, R., & Ochbelagh, D. R. (2017). Fast Neuron Detection. In M. Ahmadi, R. Ismail, & S. Anwar (Eds.), *Handbook of Research on Nanoelectronic Sensor Modeling and Applications* (pp. 395–422). Hershey, PA: IGI Global. doi:10.4018/978-1-5225-0736-9.ch015

Khalaj, Z., Monajjemi, M., & Diudea, M. V. (2017). Main Allotropes of Carbon: A Brief Review. In M. Putz & M. Mirica (Eds.), *Sustainable Nanosystems Development, Properties, and Applications* (pp. 185–213). Hershey, PA: IGI Global. doi:10.4018/978-1-5225-0492-4.ch006

Khanday, M. F. (2015). Convergence of Nanotechnology and Microbiology. In M. Shah, M. Bhat, & J. Davim (Eds.), *Nanotechnology Applications for Improvements in Energy Efficiency and Environmental Management* (pp. 313–342). Hershey, PA: IGI Global. doi:10.4018/978-1-4666-6304-6.ch011

Khayyat, M. M., & Aïssa, B. (2014). Si-NWs: Major Advances in Synthesis and Applications. In M. Bououdina & J. Davim (Eds.), *Handbook of Research on Nanoscience, Nanotechnology, and Advanced Materials* (pp. 108–130). Hershey, PA: IGI Global. doi:10.4018/978-1-4666-5824-0.ch005

Kiani, M. J., Abadi, M. H., Rahmani, M., Ahmadi, M. T., Harun, F. K., & Bagherifard, K. (2017). Graphene Based-Biosensor: Graphene Based Electrolyte Gated Graphene Field Effect Transistor. In M. Ahmadi, R. Ismail, & S. Anwar (Eds.), *Handbook of Research on Nanoelectronic Sensor Modeling and Applications* (pp. 265–293). Hershey, PA: IGI Global. doi:10.4018/978-1-5225-0736-9.ch011

Kiani, M. J., Abadi, M. H., Rahmani, M., Ahmadi, M. T., Harun, F. K., Hedayat, S., & Yaghoobian, S. (2017). Carbon Materials Based Ion Sensitive Field Effect Transistor (ISFET): The Emerging Potentials of Nanostructured Carbon-Based ISFET with High Sensitivity. In M. Ahmadi, R. Ismail, & S. Anwar (Eds.), *Handbook of Research on Nanoelectronic Sensor Modeling and Applications* (pp. 334–360). Hershey, PA: IGI Global. doi:10.4018/978-1-5225-0736-9.ch013

Krasnoholovets, V. (2017). A Theoretical Study of the Refractive Index of KDP Crystal Doped with TiO2 Nanoparticles. In M. Putz & M. Mirica (Eds.), *Sustainable Nanosystems Development, Properties, and Applications* (pp. 524–534). Hershey, PA: IGI Global. doi:10.4018/978-1-5225-0492-4.ch013

Krishnapriya, K., & Ramesh, M. (2017). Copper and Copper Nanoparticles Induced Hematological Changes in a Freshwater Fish Labeo rohita – A Comparative Study: Copper and Copper Nanoparticle Toxicity to Fish. In S. Joo (Ed.), *Applying Nanotechnology for Environmental Sustainability* (pp. 352–375). Hershey, PA: IGI Global. doi:10.4018/978-1-5225-0585-3.ch015

Kvasnička, V., & Pospíchal, J. (2014). A Study of Replicators and Hypercycles by Hofstadters Typogenetics. *International Journal of Signs and Semiotic Systems, 3*(1), 10–26. doi:10.4018/ijsss.2014010102

Kwon, K., Li, L., & Kim, D. (2015). Energy Harvesting from Wastewater Using Nanofluidic Reverse Electrodialysis. In L. Mescia, O. Losito, & F. Prudenzano (Eds.), *Innovative Materials and Systems for Energy Harvesting Applications* (pp. 380–411). Hershey, PA: IGI Global. doi:10.4018/978-1-4666-8254-2.ch013

László, I., Zsoldos, I., & Fülep, D. (2017). Self Organizing Carbon Structures: Tight Binding Molecular Dynamics Calculations. In M. Putz & M. Mirica (Eds.), *Sustainable Nanosystems Development, Properties, and Applications* (pp. 46–58). Hershey, PA: IGI Global. doi:10.4018/978-1-5225-0492-4.ch002

Lavanya, K., Durai, M. S., & Iyengar, N. (2016). A Hybrid Model for Rice Disease Diagnosis Using Entropy Based Neuro Genetic Algorithm. *International Journal of Agricultural and Environmental Information Systems*, *7*(2), 52–69. doi:10.4018/IJAEIS.2016040103

Li, H., & Harruna, I. (2014). Functionalization of Carbon Nanocomposites with Ruthenium Bipyridine and Terpyridine Complex. In M. Bououdina & J. Davim (Eds.), *Handbook of Research on Nanoscience, Nanotechnology, and Advanced Materials* (pp. 26–61). Hershey, PA: IGI Global. doi:10.4018/978-1-4666-5824-0.ch002

Li, K., & Kobayashi, T. (2017). Ionic Liquids and Poly (Ionic Liquid)s Used as Green Solvent and Ultrasound Responded Materials. In T. Kobayashi (Ed.), *Applied Environmental Materials Science for Sustainability* (pp. 327–346). Hershey, PA: IGI Global. doi:10.4018/978-1-5225-1971-3.ch015

Li, M. (2014). Pharmacokinetics of Polymeric Nanoparticles at Whole Body, Organ, Cell, and Molecule Levels. In M. Bououdina & J. Davim (Eds.), *Handbook of Research on Nanoscience, Nanotechnology, and Advanced Materials* (pp. 146–163). Hershey, PA: IGI Global. doi:10.4018/978-1-4666-5824-0.ch007

López, C. Y., Bueno, J. J., Torres, I. Z., Mendoza-López, M. L., Álvarez, J. E., & Macías, A. H. (2015). Electrophoretical Deposition of Nanotube TiO2 Conglomerates Detached During Ti Anodizing Used for Decomposing Methyl Orange in Water. In S. Soni, A. Salhotra, & M. Suar (Eds.), *Handbook of Research on Diverse Applications of Nanotechnology in Biomedicine, Chemistry, and Engineering* (pp. 477–495). Hershey, PA: IGI Global. doi:10.4018/978-1-4666-6363-3.ch022

Lucas, C. E., & Pueyo, C. L. (2016). Single Source Precursors for Semiconducting Metal Oxide-Based Films. In A. Zuzuarregui & M. Morant-Miñana (Eds.), *Research Perspectives on Functional Micro- and Nanoscale Coatings* (pp. 26–41). Hershey, PA: IGI Global. doi:10.4018/978-1-5225-0066-7.ch002

Marcu, I., Urdă, A., Popescu, I., & Hulea, V. (2017). Layered Double Hydroxides-Based Materials as Oxidation Catalysts. In M. Putz & M. Mirica (Eds.), *Sustainable Nanosystems Development, Properties, and Applications* (pp. 59–121). Hershey, PA: IGI Global. doi:10.4018/978-1-5225-0492-4.ch003

Martin, A., Kalevaru, V. N., & Radnik, J. (2016). Palladium in Heterogeneous Oxidation Catalysis. In H. Al-Megren & T. Xiao (Eds.), *Petrochemical Catalyst Materials, Processes, and Emerging Technologies* (pp. 53–81). Hershey, PA: IGI Global. doi:10.4018/978-1-4666-9975-5.ch003

Matanovic, D. (2014). General Approach to Risk Analysis. In D. Matanovic, N. Gaurina-Medjimurec, & K. Simon (Eds.), *Risk Analysis for Prevention of Hazardous Situations in Petroleum and Natural Gas Engineering* (pp. 1–22). Hershey, PA: IGI Global. doi:10.4018/978-1-4666-4777-0.ch001

McDonald, K. D., Ojo, E. O., & Liebman, J. F. (2017). What Are the Structures of the Octet Rule Obeying All-Carbon Species Cx ($2 \leq x \leq 7$ and Larger x)?: A Pedagogical, Mathematical, and Pictorial Study. In M. Putz & M. Mirica (Eds.), *Sustainable Nanosystems Development, Properties, and Applications* (pp. 1–45). Hershey, PA: IGI Global. doi:10.4018/978-1-5225-0492-4.ch001

Melnyczuk, J. M., & Palchoudhury, S. (2014). Synthesis and Characterization of Iron Oxide Nanoparticles. In M. Bououdina & J. Davim (Eds.), *Handbook of Research on Nanoscience, Nanotechnology, and Advanced Materials* (pp. 89–107). Hershey, PA: IGI Global. doi:10.4018/978-1-4666-5824-0.ch004

Meshginqalam, B., Ahmadi, M. T., Tousi, H. T., Sabatyan, A., & Centeno, A. (2017). Surface Plasmon Resonance-Based Sensor Modeling. In M. Ahmadi, R. Ismail, & S. Anwar (Eds.), *Handbook of Research on Nanoelectronic Sensor Modeling and Applications* (pp. 361–394). Hershey, PA: IGI Global. doi:10.4018/978-1-5225-0736-9.ch014

Mir, S. A., Shah, M. A., Mir, M. M., & Iqbal, U. (2017). New Horizons of Nanotechnology in Agriculture and Food Processing Industry. In B. Nayak, A. Nanda, & M. Bhat (Eds.), *Integrating Biologically-Inspired Nanotechnology into Medical Practice* (pp. 230–258). Hershey, PA: IGI Global. doi:10.4018/978-1-5225-0610-2.ch009

Mischler, S., & Roy, M. (2015). Tribocorrosion of Thermal Sprayed Coatings. In M. Roy & J. Davim (Eds.), *Thermal Sprayed Coatings and their Tribological Performances* (pp. 25–60). Hershey, PA: IGI Global. doi:10.4018/978-1-4666-7489-9.ch002

Mishra, G., Pandey, S., Dutta, A., & Giri, K. (2017). Nanotechnology Applications for Sustainable Crop Production. In S. Joo (Ed.), *Applying Nanotechnology for Environmental Sustainability* (pp. 164–184). Hershey, PA: IGI Global. doi:10.4018/978-1-5225-0585-3.ch008

Misra, R., & Rao, N. N. (2015). Electrochemical Technologies for Industrial Effluent Treatment. In N. Gaurina-Medjimurec (Ed.), *Handbook of Research on Advancements in Environmental Engineering* (pp. 118–146). Hershey, PA: IGI Global. doi:10.4018/978-1-4666-7336-6.ch005

Mitra-Kirtley, S., Mullins, O. C., & Pomerantz, A. E. (2016). Sulfur and Nitrogen Chemical Speciation in Crude Oils and Related Carbonaceous Materials. In T. Saleh (Ed.), *Applying Nanotechnology to the Desulfurization Process in Petroleum Engineering* (pp. 53–83). Hershey, PA: IGI Global. doi:10.4018/978-1-4666-9545-0.ch002

Mohamed, A. M., Abdullah, A. M., Al-Maadeed, M., & Bahgat, A. (2016). Fundamental, Fabrication and Applications of Superhydrophobic Surfaces. In A. Zuzuarregui & M. Morant-Miñana (Eds.), *Research Perspectives on Functional Micro- and Nanoscale Coatings* (pp. 341–368). Hershey, PA: IGI Global. doi:10.4018/978-1-5225-0066-7.ch013

Montalvan-Sorrosa, D., de los Cobos-Vasconcelos, D., & Gonzalez-Sanchez, A. (2016). Nanotechnology Applied to the Biodesulfurization of Fossil Fuels and Spent Caustic Streams. In T. Saleh (Ed.), *Applying Nanotechnology to the Desulfurization Process in Petroleum Engineering* (pp. 378–389). Hershey, PA: IGI Global. doi:10.4018/978-1-4666-9545-0.ch012

Moor, K., Snow, S., & Kim, J. (2017). Light Sensitized Disinfection with Fullerene. In S. Joo (Ed.), *Applying Nanotechnology for Environmental Sustainability* (pp. 137–163). Hershey, PA: IGI Global. doi:10.4018/978-1-5225-0585-3.ch007

Nogueira, V. I., Gavina, A., Bouguerra, S., Andreani, T., Lopes, I., Rocha-Santos, T., & Pereira, R. (2017). Ecotoxicity and Toxicity of Nanomaterials with Potential for Wastewater Treatment Applications. In S. Joo (Ed.), *Applying Nanotechnology for Environmental Sustainability* (pp. 294–329). Hershey, PA: IGI Global. doi:10.4018/978-1-5225-0585-3.ch013

Nunnelley, K. G., & Smith, J. A. (2017). Nanotechnology for Filtration-Based Point-of-Use Water Treatment: A Review of Current Understanding. In S. Joo (Ed.), *Applying Nanotechnology for Environmental Sustainability* (pp. 27–49). Hershey, PA: IGI Global. doi:10.4018/978-1-5225-0585-3.ch002

Obayya, S., Areed, N. F., Hameed, M. F., & Abdelrazik, M. H. (2015). Optical Nano-Antennas for Energy Harvesting. In L. Mescia, O. Losito, & F. Prudenzano (Eds.), *Innovative Materials and Systems for Energy Harvesting Applications* (pp. 26–62). Hershey, PA: IGI Global. doi:10.4018/978-1-4666-8254-2.ch002

Ogunlaja, A. S., & Tshentu, Z. R. (2016). Molecularly Imprinted Polymer Nanofibers for Adsorptive Desulfurization. In T. Saleh (Ed.), *Applying Nanotechnology to the Desulfurization Process in Petroleum Engineering* (pp. 281–336). Hershey, PA: IGI Global. doi:10.4018/978-1-4666-9545-0.ch010

Pakseresht, A., Rahimipour, M., Alizadeh, M., Hadavi, S., & Shahbazkhan, A. (2016). Concept of Advanced Thermal Barrier Functional Coatings in High Temperature Engineering Components. In A. Zuzuarregui & M. Morant-Miñana (Eds.), *Research Perspectives on Functional Micro- and Nanoscale Coatings* (pp. 396–419). Hershey, PA: IGI Global. doi:10.4018/978-1-5225-0066-7.ch015

Pau, J. L., Marín, A. G., Hernández, M. J., Cervera, M., & Piqueras, J. (2016). Analysis of Plasmonic Structures by Spectroscopic Ellipsometry. In A. Zuzuarregui & M. Morant-Miñana (Eds.), *Research Perspectives on Functional Micro- and Nanoscale Coatings* (pp. 208–239). Hershey, PA: IGI Global. doi:10.4018/978-1-5225-0066-7.ch008

Penchovsky, R. (2014). Nucleic Acids-Based Nanotechnology: Engineering Principals and Applications. In M. Bououdina & J. Davim (Eds.), *Handbook of Research on Nanoscience, Nanotechnology, and Advanced Materials* (pp. 414–430). Hershey, PA: IGI Global. doi:10.4018/978-1-4666-5824-0.ch016

Petrescu, L., Avram, S., Mernea, M., & Mihailescu, D. F. (2017). Up-Converting Nanoparticles: Promising Markers for Biomedical Applications. In M. Putz & M. Mirica (Eds.), *Sustainable Nanosystems Development, Properties, and Applications* (pp. 490–523). Hershey, PA: IGI Global. doi:10.4018/978-1-5225-0492-4.ch012

Philippe, A. (2017). Evaluation of Currently Available Techniques for Studying Colloids in Environmental Media: Introduction to Environmental Nanometrology. In S. Joo (Ed.), *Applying Nanotechnology for Environmental Sustainability* (pp. 1–26). Hershey, PA: IGI Global. doi:10.4018/978-1-5225-0585-3.ch001

Pirsa, S. (2017). Chemiresistive Gas Sensors Based on Conducting Polymers. In M. Ahmadi, R. Ismail, & S. Anwar (Eds.), *Handbook of Research on Nanoelectronic Sensor Modeling and Applications* (pp. 150–180). Hershey, PA: IGI Global. doi:10.4018/978-1-5225-0736-9.ch006

Popescu, L., Robu, A. C., & Zamfir, A. D. (2017). Sustainable Nanosystem Development for Mass Spectrometry: Applications in Proteomics and Glycomics. In M. Putz & M. Mirica (Eds.), *Sustainable Nanosystems Development, Properties, and Applications* (pp. 535–568). Hershey, PA: IGI Global. doi:10.4018/978-1-5225-0492-4.ch014

Pourasl, A. H., Ahmadi, M. T., Rahmani, M., Ismail, R., & Tan, M. L. (2017). Graphene and CNT Field Effect Transistors Based Biosensor Models. In M. Ahmadi, R. Ismail, & S. Anwar (Eds.), *Handbook of Research on Nanoelectronic Sensor Modeling and Applications* (pp. 294–333). Hershey, PA: IGI Global. doi:10.4018/978-1-5225-0736-9.ch012

Rafeeqi, T. A. (2015). Carbon Nanotubes: Basics, Biocompatibility, and Bio-Applications Including Their Use as a Scaffold in Cell Culture Systems. In M. Shah, M. Bhat, & J. Davim (Eds.), *Nanotechnology Applications for Improvements in Energy Efficiency and Environmental Management* (pp. 56–86). Hershey, PA: IGI Global. doi:10.4018/978-1-4666-6304-6.ch003

Rahmani, M., Karimi, H., Kiani, M. J., Pourasl, A. H., Rahmani, K., Ahmadi, M. T., & Ismail, R. (2017). Modeling Trilayer Graphene-Based DET Characteristics for a Nanoscale Sensor. In M. Ahmadi, R. Ismail, & S. Anwar (Eds.), *Handbook of Research on Nanoelectronic Sensor Modeling and Applications* (pp. 19–38). Hershey, PA: IGI Global. doi:10.4018/978-1-5225-0736-9.ch002

Rahmani, M., Rahmani, K., Kiani, M. J., Karimi, H., Akbari, E., Ahmadi, M. T., & Ismail, R. (2017). Development of Gas Sensor Model for Detection of NO2 Molecules Adsorbed on Defect-Free and Defective Graphene. In M. Ahmadi, R. Ismail, & S. Anwar (Eds.), *Handbook of Research on Nanoelectronic Sensor Modeling and Applications* (pp. 208–223). Hershey, PA: IGI Global. doi:10.4018/978-1-5225-0736-9.ch008

Raniero, W., Della Mea, G., & Campostrini, M. (2016). Functionalization of Surfaces with Optical Coatings Produced by PVD Magnetron Sputtering. In A. Zuzuarregui & M. Morant-Miñana (Eds.), *Research Perspectives on Functional Micro- and Nanoscale Coatings* (pp. 170–207). Hershey, PA: IGI Global. doi:10.4018/978-1-5225-0066-7.ch007

Rodulfo-Baechler, S. M. (2016). Dual Role of Perovskite Hollow Fiber Membrane in the Methane Oxidation Reactions. In H. Al-Megren & T. Xiao (Eds.), *Petrochemical Catalyst Materials, Processes, and Emerging Technologies* (pp. 385–430). Hershey, PA: IGI Global. doi:10.4018/978-1-4666-9975-5.ch014

Rongione, N. A., Floerke, S. A., & Celik, E. (2017). Developments in Antibacterial Disinfection Techniques: Applications of Nanotechnology. In S. Joo (Ed.), *Applying Nanotechnology for Environmental Sustainability* (pp. 185–203). Hershey, PA: IGI Global. doi:10.4018/978-1-5225-0585-3.ch009

Roy, M. (2014). Nano Indentation Response of Various Thin Films Used for Tribological Applications. In M. Bououdina & J. Davim (Eds.), *Handbook of Research on Nanoscience, Nanotechnology, and Advanced Materials* (pp. 62–88). Hershey, PA: IGI Global. doi:10.4018/978-1-4666-5824-0.ch003

Sadeghi, H., & Sangtarash, S. (2017). Silicene Nanoribbons and Nanopores for Nanoelectronic Devices and Applications. In M. Ahmadi, R. Ismail, & S. Anwar (Eds.), *Handbook of Research on Nanoelectronic Sensor Modeling and Applications* (pp. 39–69). Hershey, PA: IGI Global. doi:10.4018/978-1-5225-0736-9.ch003

Saikia, P., Bharadwaj, S. K., & Miah, A. T. (2016). Peroxovanadates and Its Bio-Mimicking Relation with Vanadium Haloperoxidases. In M. Bououdina (Ed.), *Emerging Research on Bioinspired Materials Engineering* (pp. 199–221). Hershey, PA: IGI Global. doi:10.4018/978-1-4666-9811-6.ch007

Saladino, R., Botta, G., & Crucianelli, M. (2016). Advances in Nanotechnology Transition Metal Catalysts in Oxidative Desulfurization (ODS) Processes: Nanotechnology Applied to ODS Processing. In T. Saleh (Ed.), *Applying Nanotechnology to the Desulfurization Process in Petroleum Engineering* (pp. 180–215). Hershey, PA: IGI Global. doi:10.4018/978-1-4666-9545-0.ch006

Saleh, T. A., Danmaliki, G. I., & Shuaib, T. D. (2016). Nanocomposites and Hybrid Materials for Adsorptive Desulfurization. In T. Saleh (Ed.), *Applying Nanotechnology to the Desulfurization Process in Petroleum Engineering* (pp. 129–153). Hershey, PA: IGI Global. doi:10.4018/978-1-4666-9545-0.ch004

Saleh, T. A., Shuaib, T. D., Danmaliki, G. I., & Al-Daous, M. A. (2016). Carbon-Based Nanomaterials for Desulfurization: Classification, Preparation, and Evaluation. In T. Saleh (Ed.), *Applying Nanotechnology to the Desulfurization Process in Petroleum Engineering* (pp. 154–179). Hershey, PA: IGI Global. doi:10.4018/978-1-4666-9545-0.ch005

Salehi, M. A., Kadusarai, M. J., & Dogolsar, M. A. (2016). Capability of Bacterial Cellulose Membrane in Release of Doxycycline. *International Journal of Chemoinformatics and Chemical Engineering*, *5*(1), 44–55. doi:10.4018/IJCCE.2016010104

Sareen, N., & Bhattacharya, S. (2016). Cleaner Energy Fuels: Hydrodesulfurization and Beyond. In T. Saleh (Ed.), *Applying Nanotechnology to the Desulfurization Process in Petroleum Engineering* (pp. 84–128). Hershey, PA: IGI Global. doi:10.4018/978-1-4666-9545-0.ch003

Shah, K. A., & Shah, M. A. (2014). Principles of Raman Scattering in Carbon Nanotubes. In M. Bououdina & J. Davim (Eds.), *Handbook of Research on Nanoscience, Nanotechnology, and Advanced Materials* (pp. 131–145). Hershey, PA: IGI Global. doi:10.4018/978-1-4666-5824-0.ch006

Shakir, I., Ali, Z., Rana, U. A., Nafady, A., Sarfraz, M., Al-Nashef, I., & Kang, D. et al. (2014). Nanostructured Materials for the Realization of Electrochemical Energy Storage and Conversion Devices: Status and Prospects. In M. Bououdina & J. Davim (Eds.), *Handbook of Research on Nanoscience, Nanotechnology, and Advanced Materials* (pp. 376–413). Hershey, PA: IGI Global. doi:10.4018/978-1-4666-5824-0.ch015

Sharma, P., Hussain, N., Das, M. R., Deshmukh, A. B., Shelke, M. V., Szunerits, S., & Boukherroub, R. (2014). Metal Oxide-Graphene Nanocomposites: Synthesis to Applications. In M. Bououdina & J. Davim (Eds.), *Handbook of Research on Nanoscience, Nanotechnology, and Advanced Materials* (pp. 196–225). Hershey, PA: IGI Global. doi:10.4018/978-1-4666-5824-0.ch010

Sharma, P., Hussain, N., Das, M. R., Deshmukh, A. B., Shelke, M. V., Szunerits, S., & Boukherroub, R. (2014). Metal Oxide-Graphene Nanocomposites: Synthesis to Applications. In M. Bououdina & J. Davim (Eds.), *Handbook of Research on Nanoscience, Nanotechnology, and Advanced Materials* (pp. 196–225). Hershey, PA: IGI Global. doi:10.4018/978-1-4666-5824-0.ch010

Shukla, R., Anapagaddi, R., Singh, A. K., Allen, J. K., Panchal, J. H., & Mistree, F. (2016). Integrated Computational Materials Engineering for Determining the Set Points of Unit Operations for Production of a Steel Product Mix. In S. Datta & J. Davim (Eds.), *Computational Approaches to Materials Design: Theoretical and Practical Aspects* (pp. 163–191). Hershey, PA: IGI Global. doi:10.4018/978-1-5225-0290-6.ch006

Souier, T. (2014). Conductive Probe Microscopy Investigation of Electrical and Charge Transport in Advanced Carbon Nanotubes and Nanofibers-Polymer Nanocomposites. In M. Bououdina & J. Davim (Eds.), *Handbook of Research on Nanoscience, Nanotechnology, and Advanced Materials* (pp. 343–375). Hershey, PA: IGI Global. doi:10.4018/978-1-4666-5824-0.ch014

Su, C., Puls, R. W., Krug, T. A., Watling, M. T., O'Hara, S. K., Quinn, J. W., & Ruiz, N. E. (2017). Long-Term Performance Evaluation of Groundwater Chlorinated Solvents Remediation Using Nanoscale Emulsified Zerovalent Iron at a Superfund Site. In S. Joo (Ed.), *Applying Nanotechnology for Environmental Sustainability* (pp. 92–111). Hershey, PA: IGI Global. doi:10.4018/978-1-5225-0585-3.ch005

Suar, S. K., Sinha, S., Mishra, A., & Tripathy, S. K. (2015). Fabrication of Metal@SnO2 Core-Shell Nanocomposites for Gas Sensing Applications. In S. Soni, A. Salhotra, & M. Suar (Eds.), *Handbook of Research on Diverse Applications of Nanotechnology in Biomedicine, Chemistry, and Engineering* (pp. 438–451). Hershey, PA: IGI Global. doi:10.4018/978-1-4666-6363-3.ch020

Sun, J., Wan, S., Lin, J., & Wang, Y. (2016). Advances in Catalytic Conversion of Syngas to Ethanol and Higher Alcohols. In H. Al-Megren & T. Xiao (Eds.), *Petrochemical Catalyst Materials, Processes, and Emerging Technologies* (pp. 177–215). Hershey, PA: IGI Global. doi:10.4018/978-1-4666-9975-5.ch008

Taborda, J. A., & López, E. O. (2016). Research Perspectives on Functional Micro and Nano Scale Coatings: New Advances in Nanocomposite Coatings for Severe Applications. In A. Zuzuarregui & M. Morant-Miñana (Eds.), *Research Perspectives on Functional Micro- and Nanoscale Coatings* (pp. 136–169). Hershey, PA: IGI Global. doi:10.4018/978-1-5225-0066-7.ch006

Torrens, F., & Castellano, G. (2017). Graphene and Fullenene Clusters: Molecular Polarizability and Ion–Di/Graphene Associations. In M. Putz & M. Mirica (Eds.), *Sustainable Nanosystems Development, Properties, and Applications* (pp. 569–599). Hershey, PA: IGI Global. doi:10.4018/978-1-5225-0492-4.ch015

Varahalarao, V., & Nayak, B. (2017). Microbial Nanotechnology: Mycofabrication of Nanoparticles and Their Novel Applications. In B. Nayak, A. Nanda, & M. Bhat (Eds.), *Integrating Biologically-Inspired Nanotechnology into Medical Practice* (pp. 102–131). Hershey, PA: IGI Global. doi:10.4018/978-1-5225-0610-2.ch005

Vargas-Bernal, R. (2015). Performance Analysis of Interconnects Based on Carbon Nanotubes for AMS/RF IC Design. In M. Fakhfakh, E. Tlelo-Cuautle, & M. Fino (Eds.), *Performance Optimization Techniques in Analog, Mixed-Signal, and Radio-Frequency Circuit Design* (pp. 336–363). Hershey, PA: IGI Global. doi:10.4018/978-1-4666-6627-6.ch014

Vargas-Bernal, R. (2016). Advances in Functional Nanocoatings Applied in the Aerospace Industry. In A. Zuzuarregui & M. Morant-Miñana (Eds.), *Research Perspectives on Functional Micro- and Nanoscale Coatings* (pp. 318–340). Hershey, PA: IGI Global. doi:10.4018/978-1-5225-0066-7.ch012

Vargas-Bernal, R. (2017). Modeling, Design, and Applications of the Gas Sensors Based on Graphene and Carbon Nanotubes. In M. Ahmadi, R. Ismail, & S. Anwar (Eds.), *Handbook of Research on Nanoelectronic Sensor Modeling and Applications* (pp. 181–207). Hershey, PA: IGI Global. doi:10.4018/978-1-5225-0736-9.ch007

Walker, G., Bououdina, M., Guo, Z. X., & Fruchart, D. (2014). Overview on Hydrogen Absorbing Materials: Structure, Microstructure, and Physical Properties. In M. Bououdina & J. Davim (Eds.), *Handbook of Research on Nanoscience, Nanotechnology, and Advanced Materials* (pp. 312–342). Hershey, PA: IGI Global. doi:10.4018/978-1-4666-5824-0.ch013

Wang, Z., Wu, P., Lan, L., & Ji, S. (2016). Preparation, Characterization and Desulfurization of the Supported Nickel Phosphide Catalysts. In H. Al-Megren & T. Xiao (Eds.), *Petrochemical Catalyst Materials, Processes, and Emerging Technologies* (pp. 431–458). Hershey, PA: IGI Global. doi:10.4018/978-1-4666-9975-5.ch015

Wani, I. A. (2015). Nanomaterials, Novel Preparation Routes, and Characterizations. In M. Shah, M. Bhat, & J. Davim (Eds.), *Nanotechnology Applications for Improvements in Energy Efficiency and Environmental Management* (pp. 1–40). Hershey, PA: IGI Global. doi:10.4018/978-1-4666-6304-6.ch001

Wani, I. A., & Ahmad, T. (2017). Understanding Toxicity of Nanomaterials in Biological Systems. In S. Joo (Ed.), *Applying Nanotechnology for Environmental Sustainability* (pp. 403–427). Hershey, PA: IGI Global. doi:10.4018/978-1-5225-0585-3.ch017

Wojtanowicz, A. K. (2014). Risk and Remediation of Irreducible Casing Pressure at Petroleum Wells. In D. Matanovic, N. Gaurina-Medjimurec, & K. Simon (Eds.), *Risk Analysis for Prevention of Hazardous Situations in Petroleum and Natural Gas Engineering* (pp. 155–180). Hershey, PA: IGI Global. doi:10.4018/978-1-4666-4777-0.ch008

Yasar, D., & Celik, N. (2017). Assessment of Advanced Biological Solid Waste Treatment Technologies for Sustainability. In S. Joo (Ed.), *Applying Nanotechnology for Environmental Sustainability* (pp. 204–230). Hershey, PA: IGI Global. doi:10.4018/978-1-5225-0585-3.ch010

Zarras, P., Goodman, P. A., & Stenger-Smith, J. D. (2016). Functional Polymeric Coatings: Synthesis, Properties, and Applications. In A. Zuzuarregui & M. Morant-Miñana (Eds.), *Research Perspectives on Functional Micro- and Nanoscale Coatings* (pp. 78–104). Hershey, PA: IGI Global. doi:10.4018/978-1-5225-0066-7.ch004

Zhang, T., Zhang, C., Xing, J., Xu, J., Li, C., Wang, P. C., & Liang, X. (2017). Multifunctional Dendrimers for Drug Nanocarriers. In R. Keservani, A. Sharma, & R. Kesharwani (Eds.), *Novel Approaches for Drug Delivery* (pp. 245–276). Hershey, PA: IGI Global. doi:10.4018/978-1-5225-0751-2.ch010

Zhizhin, G. V., & Diudea, M. V. (2017). Space of Nanoworld. In M. Putz & M. Mirica (Eds.), *Sustainable Nanosystems Development, Properties, and Applications* (pp. 214–236). Hershey, PA: IGI Global. doi:10.4018/978-1-5225-0492-4.ch007

Zuzuarregui, A., & Morant-Miñana, M. C. (2016). Functional Coatings: A Rapidly and Continuously Developing Field. In A. Zuzuarregui & M. Morant-Miñana (Eds.), *Research Perspectives on Functional Micro- and Nanoscale Coatings* (pp. 1–25). Hershey, PA: IGI Global. doi:10.4018/978-1-5225-0066-7.ch001

Index

M

N

O

P

S

T

U

www.ingramcontent.com/pod-product-compliance
Lightning Source LLC
Chambersburg PA
CBHW082034180926
51756CB00005B/398